Exploring *The*
BUILDING BLOCKS
of
Science

Book 3

TEACHER'S MANUAL

REBECCA W. KELLER, PhD

REAL SCIENCE 4 Kids

Exploring the Building Blocks of Science Book 3 Teacher's Manual
ISBN 978-1-941181-03-4

Published by Gravitas Publications Inc.
www.realscience4kids.com
www.gravitaspublications.com

A Note From the Author

This curriculum is designed for elementary level students and provides an introduction to the scientific disciplines of chemistry, biology, physics, geology, and astronomy. *Exploring the Building Blocks of Science Book 3 Laboratory Notebook* accompanies the *Building Blocks of Science Book 3 Student Textbook*. Together, both provide students with basic science concepts needed for developing a solid framework for real science investigation. The *Laboratory Notebook* contains 44 experiments—two experiments for each chapter of the Student Textbook. These experiments allow students to further explore concepts presented in the *Student Textbook*. This teacher's manual will help you guide students through laboratory experiments designed to help students develop the skills needed for the first step in the scientific method—making good observations.

There are several sections in each chapter of the *Laboratory Notebook*. The section called *Think About It* provides questions to help students develop critical thinking skills and spark their imagination. The *Observe It* section helps students explore how to make good observations. In every chapter there is a *What Did You Discover?* section that gives the students an opportunity to summarize the observations they have made. A section called *Why?* provides a short explanation of what students may or may not have observed. And finally, in each chapter an additional experiment is presented in *Just For Fun*.

The experiments take up to 1 hour. The materials needed for each experiment are listed on the following pages and also at the beginning of each experiment.

Enjoy!

Rebecca W. Keller, PhD

Materials at a Glance

Experiment 1	Experiment 3	Experiment 4	Experiment 5	Experiment 6
pencil or pen **Optional** colored pencils	glasses or plastic cups, several measuring cup 3 bags, small paper or plastic several small rocks (5-10) Legos (handful) sand (2 handfuls) sugar (handful) salt (2 handfuls) water food coloring, several colors 1-2 white coffee filters white paper, several sheets scissors several pencils tape	Elmer's white glue, approx. 30-60 ml (⅛-¼ cup) liquid laundry starch, approx. 30-60 ml (⅛-¼ cup)* 2 plastic cups measuring cup 30 metal paperclips *Just For Fun*: non-toxic glue such as blue glue, clear glue, wood glue, glitter glue, or paste glue; approx. 30-60 ml (⅛-¼ cup) *If you are unable to find liquid laundry starch, you can use a mixture of equal parts cornstarch and borax mixed with enough water to dissolve them. Make about 30-60 ml (⅛-¼ cup) for this experiment. **Optional** food coloring	flour, 2 liters (8 cups) 1 package active dry yeast, 7 grams (¼ oz.) lukewarm water, 240 ml (1 cup) cold water, 240 ml (1 c.) sugar, 30 ml (2 Tbsp.) vegetable oil, approx. 60 ml (4 Tbsp.) salt, 5 ml (1 tsp.) butter, 120 ml (½ cup) softened double-acting baking powder, 15 ml (1 Tbsp.) milk, 360 ml (1½ cups) measuring cups measuring spoons marking pen 4 mixing bowls mixing spoon floured bread board 2 bread pans or cookie sheets refrigerator oven timer **Optional** rolling pin biscuit cutter	notebook or drawing pad with blank pages (not ruled) to make a nature journal pencil colored pencils **Optional** camera and printer tape

Experiment 2
clear plastic cups, 15 or more measuring cup measuring spoons spoon for mixing liquid soap marking pen food items (approx. 60 ml (¼ c.) each: water milk juice vegetable oil melted butter

Experiment 7	Experiment 8	Experiment 9	Experiment 10	Experiment 11
2 small houseplants of the same kind and size 2 more small houseplants of the same kind and size water measuring cup closet or cardboard box colored pencils	2-4 white carnations 1 or more other white flowers (rose, lily, etc.) 2-3 small jars food coloring water tape knife colored pencils **Optional** magnifying glass	1-2 small clear glass jars 2 or more dried beans (white, pinto, soldier, etc.) 2 or more additional dried beans (different kind) or other seeds absorbent white paper scissors knife plastic wrap clear tape rubber band water **Optional** magnifying glass	3-5 large lemons knife 3-5 copper pennies older than 1982 3-5 galvanized (zinc coated) nails LED (Radio Shack #276-30700 [as of this writing]) 4-6 pairs alligator clips* plastic coated copper wire, .6-1.2 m (2-4 feet) wire clippers small Phillips screwdriver *duct tape can be substituted for alligator clips]	2-3 rubber balloons string or thread, at least 2 meters (6 feet) cut in half scissors different materials to rub the balloon on, such as: cotton clothing silk clothing wool clothing wooden surface plaster wall metal surface leather surface

Experiment 12	Experiment 13	Experiment 14	Experiment 15	Experiment 16
lemon battery supplies (see Experiment 10) suggested test materials: Styrofoam plastic block cotton ball nickel coin metal paper clip plastic paper clip glass of water table salt, 15 ml (1 Tbsp)	two bar magnets with the poles labeled "N" and "S"	3 Styrofoam cups: 355 ml (12 oz.) size about 240 ml (1 cup) each:* sand pebbles small rocks * student-collected or purchased from a place that sells aquarium supplies 3 containers for collecting sand, pebbles, and small rocks garden trowel or small shovel pencil 1-2 measuring cups water enough dirt, pebbles, rocks, water, etc. to make a mud city **Optional** stopwatch or clock with second hand	pencil colored pencils	2 bar magnets (narrow magnets work best) small, flat-bottomed, clear plastic box (big enough for 2 magnets to fit underneath with some space around them) corn syrup iron filings, about 5 ml (1 teaspoon) (see Experiment section for how students can collect iron filings — or iron filings may be purchased: www.hometrainingtools.com) **Optional** tape 2 plastic bags for collecting iron filings

Experiment 17	Experiment 18	Experiment 19	Experiment 20	Experiment 21
seeds (student selected) a garden bed or containers and potting soil tools for tending plants herb seeds or small herb plants (student selected) This experiment is done over the course of several weeks.	student-selected materials to make a model of a galaxy, such as colored modeling clay, Styrofoam balls, tennis balls, marbles, sand, candies, etc. cardboard or poster board, .3-1 meter (1'-3') on each side **Optional** colored pencils or markers camera and printer	colored pencils a dark, moonless night sky far away from city lights **Optional** computer with internet access pictures of cities	2 bar magnets iron filings, purchased* or student collected (see Chapter 16) shallow, flat-bottomed plastic container (or a plastic box top or large plastic jar lid) corn syrup plastic wrap Jell-O or other gelatin and items to make it assorted fruit cut in pieces and/or berries **Optional** cardboard box *As of this writing, available from Home Science Tools: http://www.hometrainingtools.com Item #CH-IRON	small plastic pail that will fit in freezer water dirt small stones dry ice (available at most grocery stores) heavy gloves or oven mitts freezer **If dry ice is in a block:** safety goggles mallet or hammer grocery bag (cloth or paper)

Experiment 22
library or internet access **Optional** old toys to take apart for computer chips (1 or more)

Materials
Quantities Needed for All Experiments

Equipment	Foods	Foods (continued)

alligator clips, 4-6 pairs (duct tape can be substituted for alligator clips)
bowls, mixing, 4
bread board
bread pans or cookie sheets, 2
freezer
jars, 2-3 small clear
knife
LED (Radio Shack #276-30700 [as of this writing])
magnets, 2 bar with the poles labeled "N" and "S"
magnets, bar, 2 narrow
measuring cup, 1-2
measuring spoons
oven
pail, small plastic that will fit in freezer
refrigerator
scissors
screwdriver, small Phillips
spoon, mixing
timer
tools for tending plants
trowel, garden, or small shovel
wire clippers
 	
Optional
  biscuit cutter
  camera
  computer with internet access
  computer printer
  magnifying glass
  rolling pin
  stopwatch or clock with second hand
	baking powder, double-acting ,15 ml (1 Tbsp.)
	beans, dried, 2 or more (white, pinto, soldier, etc.)
	beans, dried, 2 or more additional different or other seeds
	butter, 120 ml (½ cup)
	corn syrup
	flour, 2 liters (8 cups)
	food coloring, several colors
	food items (approx. 60 ml (¼ c.) each:
	  water
	  milk
	  juice
	  vegetable oil
	  melted butter
	fruit, assorted, cut in pieces and/or berries
	Jell-O or other gelatin
	lemons, 6-10 large
	milk, 360 ml (1½ cups)
	salt, 25 ml (5 tsp.) + 2 handfuls
	sugar, 30 ml (2 Tbsp.) or more
	vegetable oil, approx. 60 ml (4 Tbsp.)
	water
	yeast, active dry, 1 package, 7 grams (¼ oz)	

Materials
Quantities Needed for All Experiments

Materials	Materials (continued)	Other

bags, 3 small paper or plastic
bags, 2 plastic, for collecting iron filings
balloons, 2-3 rubber
box, small, flat-bottomed, clear plastic (big enough for 2 magnets to fit underneath with some space around them)
cardboard or poster board, .3-1 meter (1'-3') on each side
coffee filters, 1-2 white
container, shallow, flat-bottomed plastic (or a plastic box top or large plastic jar lid)
containers (3) for collecting sand, pebbles, and small rocks
cotton ball
cups, plastic, clear, 17 or more
cups, plastic or glasses, several
cups, Styrofoam (3) 355 ml (12 oz.) size
dirt
dry ice (available at most grocery stores) [If dry ice is in a block: safety goggles mallet or hammer grocery bag (cloth or paper]
flowers, carnations, 2-4 white
flowers, white, 1 or more that are not carnations (rose, lily, etc.)
gloves (heavy) or oven mitts
glue, Elmer's white, approx. 30-60 ml (⅛-¼ cup)
glue, non-toxic, such as blue glue, clear glue, wood glue, glitter glue, or paste glue; approx. 30-60 ml (⅛-¼ cup)
herb seeds or small herb plants (student selected)
houseplants, small, 4 (2 each of the same kind and size)
iron filings, about 10 ml (2 teaspoons) (student-collected or purchased: www.hometrainingtools.com, Item #CH-IRON as of this writing)
	laundry starch, liquid. approx. 30-60 ml (⅛-¼ cup), or a mixture of equal parts cornstarch and borax mixed with enough water to dissolve them
	Legos (handful) nails, 3-5 galvanized (zinc coated)
	nickel (coin)
	notebook or drawing pad with blank pages (not ruled) to make a nature journal
	paper, absorbent, white
	paper, white, several sheets
	paperclips, 30 (metal)
	paperclip, plastic
	pebbles, about 240 ml (1 cup) or more*
	pen
	pen, marking
	pencils (several)
	pencils, colored
	pennies, 3-5 copper, older than 1982
	plastic wrap
	rocks, small*
	rubber band
	sand, more than 1 cup*
	seeds (student selected)
	soap, liquid
	stones, small
	string or thread, at least 2 meters (6 feet) cut in half
	Styrofoam, small piece
	tape
	tape, clear
	water
	wire, plastic coated copper, .6-1.2 m (2-4 feet)
	
	Optional
	  markers, colored
	  pictures of cities
	  box, cardboard
	  old toys to disassemble to look for computer chips (1 or more)
	
	* student-collected or purchased from a place that sells aquarium supplies
		closet or cardboard box
		dirt, pebbles, rocks, water, etc. (enough to make a mud city)
		garden bed or containers and potting soil
		library or internet access
		materials (student-selected) to make a model of a galaxy, such as colored modeling clay, Styrofoam balls, tennis balls, marbles, sand, candies, etc.
		materials to rub a balloon on, such as:
		  cotton clothing
		  silk clothing
		  wool clothing
		  wooden surface
		  plaster wall
		  metal surface
		  leather surface
		night sky, dark moonless, far away from city lights

Contents

CHEMISTRY

BIOLOGY

PHYSICS

GEOLOGY

ASTRONOMY

Experiment 1

A Day Without Science

Materials Needed

- pencil or pen
- colored pencils (optional)

Objectives

In this experiment, students will observe how science has contributed to new technologies.

The objectives of this lesson are for students to:

- Observe how people have solved problems and invented tools.
- Understand how science and technology are interrelated.

Experiment

I. Think About It

Read this section of the *Laboratory Notebook* with your students.

Have the students think about all the things they use in a day and how science was involved in developing them.

Explore open inquiry with questions such as the following:

- *What is your breakfast made of? How did it get here? How was science involved in creating your breakfast and bringing it to you?*
  (If any breakfast items are packaged, have students look on the package, note the ingredients, and think about where they came from.)

- *What is your clothing made of? How do you think your clothing was made? Was science involved?*
  (Help them look on the label.)

- *How does your bike work? Do you think there would be bikes without science? Why or why not?*

- *How does a car work? Do you think science was used to make the car and get it to run? Why or why not?*

- *What will happen if the TV is not plugged in? Do you think this involves science? Why or why not?*

II. Observe It

Read this section of the *Laboratory Notebook* with your students.

Have your students explore the ways in which their lives are connected to modern technologies. Many of the modern conveniences that people use to manage their lives are often taken for granted. Help your students look carefully at what things are made of, what types of machines they use, where plastics are used, and when electricity is used.

III. What Did You Discover?

Read this section of the *Laboratory Notebook* with your students.

Discuss any questions that might come up.

IV. Why?

Read this section of the *Laboratory Notebook* with your students.

Have a discussion about the ways in which science impacts our lives every day. Have students think about how scientific discoveries have brought about great changes in the last 100 years, 25 years, and 10 years.

Discuss any questions that may come up.

V. Just For Fun

Encourage the students to use their imagination freely to think about what life might be like 2000 years in the future. They can draw and/or write their ideas.

Experiment 2

Make It Mix!

Materials Needed

- 15 or more clear plastic cups
- measuring cup
- measuring spoons
- spoon for mixing
- liquid soap
- marking pen
- the following food items [approx. 60 ml (1/4 cup) each]:
  water
  milk
  juice
  vegetable oil
  melted butter

Objectives

In this experiment students will observe mixtures.

The objectives of this lesson are to help students understand that:

- Liquids that are similar will mix.
- Liquids that are not similar will not mix.

Experiment

I. Think About It

Read this section of the *Laboratory Notebook* with your students.

Have the students think about whether each item in the top row is "like water" or "like oil" and then check the corresponding box below the item.

Have the students answer the questions in this section. They can refer to the chart they filled out. Their answers may vary and there are no "right" answers.

II. Observe It

Read this section of the *Laboratory Notebook* with your students.

Using clear plastic cups, help the students measure at least 60 ml (1/4 cup) of the following liquids into separate cups and label the cups with a marking pen:

- water
- milk
- juice
- oil
- butter

Using additional cups as test cups, have the students start mixing the liquids together by pouring about 15 ml (1 Tbsp.) of water into about 15 ml (1 Tbsp.) of milk. As the students make a mixture, have them use a marking pen to label the cup with the name of the mixture.

Have the students observe what happens when different liquids are mixed together. Then have them record their results in the chart in the *Laboratory Notebook*. Help them identify whether the two liquids mix or don't mix. When two liquids mix, the students won't be able to tell where one liquid starts and the other ends. When they don't mix, droplets of one liquid will be visible in the other.

Make sure the students do not confuse a color change with "mixing" or "not mixing." The liquids could change colors, but should be considered "mixed" only if there are no droplets visible.

It is not necessary to test every combination. At a minimum have the students test oil and water, oil and milk, and oil and butter.

Results should be as follows:

Results of Mixing Liquids					
	Water	Milk	Juice	Oil	Butter
Water		mixed	mixed	not mixed (oil droplets visible)	not mixed (butter droplets visible)
Milk			mixed	slightly mixed	slightly mixed
Juice				not mixed (oil droplets visible)	not mixed (butter droplets visible)
Oil					mixed
Butter					

Save the mixtures for the next part of the experiment.

To help students think about what they observed and to prepare them for the next part of the experiment, ask questions such as the following:

- When you combined water, juice, and milk with each other do you think they mixed?

- If you were to add soap to any mixture of water, juice, or milk, do you think it would make any difference to how they mix?

- When you combined butter and oil with each other, did they mix?

- If you were to add soap to a combination of butter and oil, do you think it would make any difference to whether they mix?

- When would soap be needed in order to make two liquids mix together?

Observe It With Soap

Using the mixtures from the first part of the experiment, have the students add about 2.5 ml (1/2 tsp.) of liquid soap to each mixture. The students should observe that soap doesn't change the liquids that already mix (e.g., water and juice), but does make the oil "mix" a little better into water and juice. It is not necessary to have them test all the mixtures, but have them add soap to a few of the oily mixtures and at least one of the mixtures of water-like liquids.

Their results will vary, but may look as follows:

Results of Adding Soap to Mixtures

	Water	Milk	Juice	Oil	Butter
Water		mixed	mixed	somewhat mixed	somewhat mixed
Milk			mixed	somewhat mixed	somewhat mixed
Juice				somewhat mixed	somewhat mixed
Oil					mixed
Butter					

III. What Did You Discover?

Read this section of the *Laboratory Notebook* with your students.

Help the students answer the questions in this section of the *Laboratory Notebook*. They should have observed that oil and butter do not mix with either water or juice. They also should have observed that oil mixes somewhat with milk and more with butter.

After adding soap, the students should have observed that oil mixes a little better with water and juice and much better with milk and butter.

CHEMISTRY

CHEMISTRY

IV. Why?

Read this section of the *Laboratory Notebook* with your students.

Have a discussion about the concepts presented in this section of the *Laboratory Notebook*. Explain to the students that "similar" liquids mix well, while liquids that are not "similar" do not mix well. Juice is similar to water because juice is mostly water, so juice and water mix well. Milk is a colloid, but will still mix well with water and juice, because milk is mostly water. (A colloid is a mixture that has very small droplets of molecules that do not actually mix well, but the droplets are so small it looks mixed. Colloids are often opaque). Oil and butter are similar because both oil and butter are fats. Oil and water are not similar, so oil will not mix well with either water or juice.

Explain the "rule" that similar liquids mix and dissimilar liquids do not mix.

Explain that soap is both a little bit like water and a little bit like oil, so soap mixes in both types of liquids. Because soap is like both water and oil, it "dissolves" oil in water. This is why soap works as a cleaner.

V. Just For Fun

Students are to create their own experiment to find out if certain liquids are "like water" or "like oil." Have them think of ways they could test the liquids. For example, they might choose water for the fifth liquid and stir each of the other liquids into the water to see if they mix. Allow them to try their idea even if you know it won't work. Experiments that don't work can provide valuable information for scientific research.

Substitutions can be made for items on the list. Students can look around the kitchen to see what's available. Have them think about whether they need to do anything different to test thick items like mayonnaise.

Have the students give their experiment a name and make a chart to record their observations. They can fill in the chart provided or create their own. Have them make notes about their idea for the experiment, how they performed the experiment, and how well they think their idea worked.

Experiment 3

Make It Un-mix

Materials Needed

- several glasses or plastic cups
- measuring cup
- 3 bags (small paper or plastic)
- several small rocks (5-10)
- Legos (handful)
- sand (2 handfuls)
- sugar (handful)
- salt (2 handfuls)
- water
- food coloring, several colors
- 1-2 white coffee filters
- white paper, several sheets
- scissors
- several pencils
- tape

Objectives

In this experiment students will explore techniques used for separating various mixtures.

The objectives of this lesson are to have students:

- Gain a basic understanding of mixtures and the separation of mixtures.
- Explore different ways of separating mixtures of large, dissimilar items.
- Find ways to separate mixtures that have small, similar components.
- Experiment with a technique called chromatography that can be used to separate molecules from mixtures.

Experiment

I. Think About It

Read this section of the *Laboratory Notebook* with your students and discuss the questions with them. Help them think of things they might do to separate several different kinds of mixtures. Their answers may vary. Encourage them to think of different "tools," such as a sieve or flour sifter for separating mixtures. Also guide them to think of using water to dissolve part of a mixture, such as salt in the salt/sand mixture. There are no right answers to these questions.

II. Observe It

Read this section of the *Laboratory Notebook* with your students.

Have the students test one or more of their own ideas for separating each mixture. Even if you know their idea won't work, let them test it. Answers will vary—possible answers follow.

❶ Take a handful of rocks and a handful of Legos and mix them together on the table. Now try to un-mix them. Draw or describe what you did.

used hands and fingers to un-mix the rocks from the Legos

used a cardboard box with holes in it to un-mix the rocks from the Legos

(Answers may vary.)

Rocks and Legos

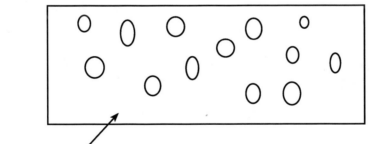

cardboard box with holes in it

❷ Take a handful of rocks and mix them with sand in a bag. Now un-mix the rocks and sand. Draw or describe what you did.

used a sieve to separate rocks and sand

used a hair dryer to blow away all of the sand

used cheesecloth to separate rocks and sand

(Answers may vary.)

Rocks and Sand

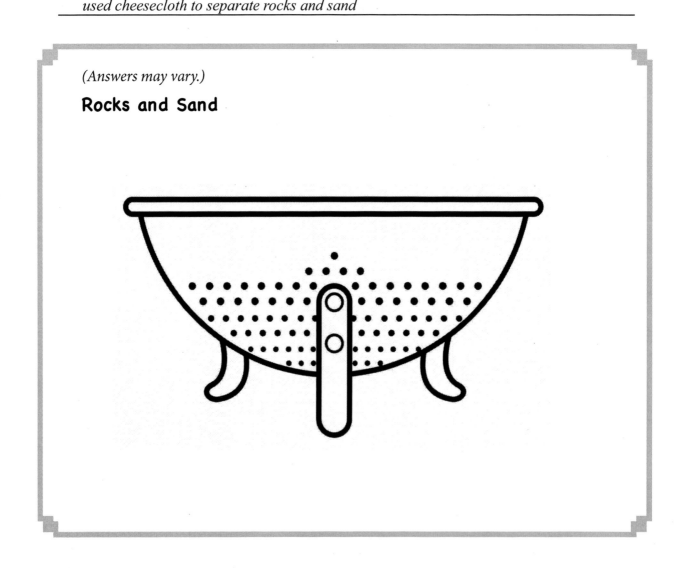

❸ Take a handful of sand and a handful of salt and mix them in a bag. Now un-mix them. Draw or describe what you did.

used water to dissolve the salt

_____

_____

_____

_____

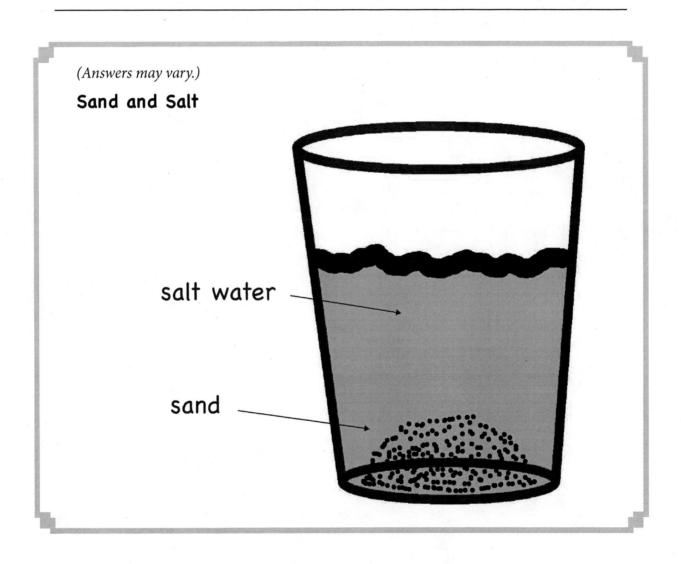

(Answers may vary.)

Sand and Salt

salt water

sand

Paper Chromatography

The next steps in the experiment involve mixing different colors of food coloring and then separating the colors.

❹ Have the students place several drops of different colors of food coloring into 120 ml (1/2 cup) of water. The resulting color should be black or deep brown. Discuss possible ways to separate the colors.

❺ Next, discuss a method called *chromatography* that can be used to separate the colors from the water and from each other. Explain to the students that they can separate the colors by using a piece of coffee filter paper.

Help the students set up the chromatography sample. Have them cut the filter paper into long strips, tape one end of one strip to a pencil, and place the pencil across the glass containing the colored water, letting the paper strip dip into the water.

❻ They should observe the water immediately begin to migrate up the paper strip. They should detect the green food coloring migrating first, followed by the blue, then yellow, and finally red.

When they take the paper strip out of the water, have them lay it down on a piece of white paper. They should easily see the different colors. Have them record their results. When the paper strip is dry, it can be taped in the box in the *Laboratory Notebook*.

❼ Have them repeat the experiment with an "unknown." Without the students observing, add several drops of two or three colors into 120 ml (1/2 cup) of water in a glass. Give the glass to the students, and let them perform paper chromatography to determine which colors are in the water.

❽ Have the students prepare an "unknown" for the teacher, and let the teacher separate the colors. This is a lot of fun and can be repeated as many times as you wish.

❾ Have them record the results for Steps ❼ and ❽.

III. What Did You Discover?

Read this section of the *Laboratory Notebook* with your students.

Have the students answer the questions in this section of the *Laboratory Workbook*. Their answers will vary.

IV. Why?

Read this section of the *Laboratory Notebook* with your students.

Lead a discussion of the concepts covered in this section. Explain that there are many different ways to separate mixtures. Review the different ways the students discovered to separate the mixtures in their experiment.

Also discuss why some mixtures are easier to separate than others. Mixtures that have small components and mixtures that are made of similar items are harder to separate than mixtures with larger components and dissimilar items. Explain that scientists use a variety of tools and techniques to separate mixtures. The "trick" called chromatography is a technique frequently used by scientists to separate a variety of molecules. Explain that chromatography can be used to separate different kinds of molecules, such as proteins or DNA, and not just molecules that make color.

V. Just For Fun

CHEMISTRY

Take a handful of salt and a handful of sugar and mix them in a bag. Now un-mix them. Draw or describe what you did.

(*Answers may vary.*)

Salt and Sugar

[This one is just for fun. Encourage the students to use their imagination.]

Used ants at a picnic

Since ants like sugar better than salt, the ants will carry off the sugar and leave the salt.

Experiment 4

Making Goo

Materials Needed

- Elmer's white glue, approx. 30-60 ml (1/8-1/4 cup)
- liquid laundry starch, approx. 30-60 ml (1/8-1/4 cup)*
- measuring cup
- 2 plastic cups
- 30 metal paperclips
- *Just For Fun* section: non-toxic glue such as blue glue, clear glue, wood glue, glitter glue, or paste glue, approx. 30-60 ml (1/8-1/4 cup)

*If you are unable to find liquid laundry starch, you can use a mixture of equal parts cornstarch and borax mixed with enough water to dissolve them. Make about 30-60 ml (1/8-1/4 cup) for this experiment.

Optional

- food coloring

Objectives

In this experiment students will explore the chemical reaction between Elmer's glue and liquid laundry starch.

The objectives of this lesson are to:

- Introduce the concept of polymers.
- Have students observe a chemical reaction that changes the properties of two polymers by the formation of cross-links.

Experiment

I. Think About It

Read this section of the *Laboratory Notebook* with your students.

Have the students think about the questions in this section of the *Laboratory Notebook* and encourage them to explore their ideas freely. Their answers may vary from those you expect. There are no right answers.

Guide inquiry with questions such as the following:

- *When you use glue, what does it feel like if you get it on your fingers?*

- *If you put your fingers in laundry starch (or cornstarch/borax mixed with water), what does it feel like?*

- *If you mixed glue and laundry starch together, do you think they would stay the same or change? Why or why not?*

- *What do you think the mixture would be like? Sticky? Slippery? Watery? Thick? What else might it be like?*

II. Observe It

Read this section of the *Laboratory Notebook* with your students.

In this experiment students will be adding liquid laundry starch or a cornstarch/borax/water mixture to Elmer's glue. Have the students measure 30-60 ml (⅛-¼ cup) of glue and pour it into a plastic cup. It is important not to put too much glue in the cup since students may need to add more laundry starch (or cornstarch/borax) than glue.

Have the students add 30-60 ml (⅛-¼ cup) laundry starch (or cornstarch/borax) to the glue. Nothing will happen until the students knead the glue and starch mixture. Have them add more laundry starch (or cornstarch/borax) if necessary. Encourage students to knead the mixture with their fingers. This is messy for teachers but delightful for most students. Both the glue and starch are nontoxic and can be easily cleaned from clothing and hands.

As students knead the mixture, help them think about what changes they may be observing. They should feel the glue become less sticky and more rubbery. They should be able to get the glue to roll into a ball or flatten in their hands like a pancake.

III. What Did You Discover?

Read this section of the *Laboratory Notebook* with your students.

Have the students answer the questions by describing what they actually observed. They should have noticed a significant change in the properties of the glue when it was kneaded with the starch.

IV. Why?

Read this section of the *Laboratory Notebook* with your students.

The glue and the laundry starch (or cornstarch/borax mixture) are both *polymers*, which are long chains made of hooked together molecules. When these two polymers are kneaded together, a chemical reaction occurs, changing the properties of the polymers. In this case, the starch makes the long chains of molecules in the glue hook to each other. This is called cross-linking because it makes cross-links between the polymers.

To illustrate this principle, gather 30 paperclips. Have the students make three chains with 10 paperclips each. Have the students lay them side by side on the table. Show the students that they can slide the chains of paperclips past each other. Explain that this is how the glue behaves without the laundry starch. Now take the paperclips and hook two or three from one chain to two or three from another chain. This is a cross-link. Next show the students that they can no longer easily slide the chains back and forth with respect to each other. This illustrates the changes that occur when the starch is kneaded into the glue.

Slides easily : no cross-links

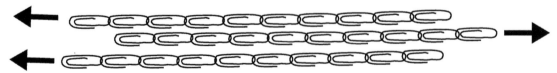

No sliding

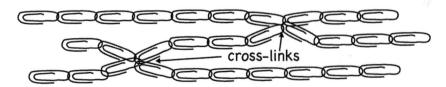

cross-links

V. Just For Fun

Students are to repeat the experiment using a different glue to see if this substitution changes the results. Help the students select a non-toxic glue.

If the students used liquid laundry starch in the original experiment, you can have them substitute the borax/cornstarch mixture for the laundry starch instead of using a different glue.

Students can add a few drops of food coloring to their mixture to see what it will look like.

Have them record their results.

CHEMISTRY

Experiment 5

Make It Rise!

Materials Needed

- flour, 2 liters (8 cups)
- 1 package active dry yeast, 7 grams (.25 oz.)
- lukewarm water, 240 ml (1 cup)
- cold water, 240 ml (1 cup)
- sugar, 30 ml (2 Tbsp.)
- vegetable oil
- 5 ml (1 tsp.) salt
- 120 ml (1/2 cup) soft butter
- 15 ml (1 Tbsp.) double-acting baking powder
- 360 ml (1 1/2 cups) milk
- measuring cups
- measuring spoons
- 4 mixing bowls
- mixing spoon
- floured bread board
- 2 bread pans or cookie sheets
- 2 cookie sheets
- marking pen
- refrigerator
- oven
- timer

Optional

- rolling pin
- biscuit cutter

Objectives

In this experiment students will observe how different temperatures affect the activity of enzymes in yeast.

The objectives of this lesson are:

- To introduce the concept of enzymes.
- To have students observe that enzymes must be within a certain temperature range in order to function properly.

Experiment

I. Think About It

Read this section of the *Laboratory Notebook* with your students.

Have the students answer the questions in this section of the *Laboratory Workbook*. Their answers will vary. Sample answers are given.

❶ List as many different types of molecules as you can.

acids, bases, salt, sugar, oils, water, enzymes

❷ What kinds of molecules make food salty?

salt molecules

❸ What kinds of molecules are glue and starch made of?

long chains (polymers)

❹ Do you think all the molecules in your body would work properly if your body got too hot? Why or why not?

(answers will vary)

❺ What kinds of molecules do you think make bread rise?

salt molecules? sugar molecules? acid molecules?

II. Observe It

Read this section of the *Laboratory Notebook* with your students.

Using the directions in the *Laboratory Notebook,* help the students make two rounds of bread dough. One dough will be made with warm water and placed in a warm place to rise—**Dough A**. The other dough will be made using cold water and placed in the refrigerator to rise—**Dough B**. The students should observe that **Dough A** rises and **Dough B** does not rise. Make sure that the warm water added to **Dough A** is not too hot. Hot water will kill the yeast.

III. What Did You Discover?

Read this section of the *Laboratory Notebook* with your students and have the students answer the questions.

Students should observe a significant difference between **Dough A** and **Dough B.** Help the students find words to describe what happened to the two doughs and what was different. Help them connect the fact that one dough was made with cold water and kept in a cold place, and the other dough was made with warm water and kept in a warm place. Point out to the students that temperature was the only difference between the two doughs. Help them see that this one change was what caused one dough to rise and the other dough not to rise.

IV. Why?

Read this section of the *Laboratory Notebook* with your students and discuss the concepts presented. Help the students understand that yeast is a living thing that contains very large protein molecules called *enzymes* that produce the gases needed for yeast to make bread rise. Yeast contains enzymes that convert sugar to carbon dioxide and alcohol. This is called fermentation. The carbon dioxide gas that is produced during the fermentation process is what makes the bread rise. The alcohol is burned off during the baking process.

Explain that there are many different kinds enzymes that perform a variety of tasks. Each enzyme is a large, complicated molecule that is shaped in a particular way and designed to perform a particular function. There are enzymes that cut molecules, enzymes that copy molecules, enzymes that glue molecules together, and enzymes that read other molecules.

Explain to the students that there are enzymes in their body that can only function within a narrow temperature range. The enzymes cannot function properly if the body temperature is either too hot or too cold.

V. Just For Fun

In this experiment students will observe what happens when biscuits are made with and without baking powder.

Baking Powder (or Not) Biscuits

Have the students think about what will happen when they make biscuits with and without baking powder. Have them record their ideas in the box provided.

Using the directions in the *Laboratory Notebook,* help the students make two biscuit dough mixtures. **Dough A** will be made with baking powder. **Dough B** will be made without baking powder. When they put the biscuits on cookie sheets, help them keep track of which cookie sheet contains **Dough A** and which has **Dough B**. The students should observe that, when they are baked, **Dough A** rises and **Dough B** does not rise.

Students should observe a significant difference between **Dough A** and **Dough B**. Have the students describe what happened to the two doughs and what was different. Point out to the students that adding baking powder or not adding it was the only difference between the two doughs. Help them see that this one change was what caused one dough to rise and the other dough not to rise.

Baking powder contains an acid and a base (often sodium bicarbonate, or baking soda) and when it gets wet, a chemical reaction occurs that releases bubbles of carbon dioxide. In recipes where baking soda is used, the recipe will call for the addition of an acid ingredient in liquid form (such as buttermilk or vinegar) to react with the baking soda.

Explain to the students that bread made with yeast and baking powder biscuits both rise due to release of carbon dioxide gas. However, the carbon dioxide in the bread is made by enzymes in yeast, a living thing, and the carbon dioxide from baking powder is made by a chemical reaction of non-living substances.

Experiment 6

Nature Walk

Materials Needed

- notebook or drawing pad with blank pages (not ruled) to make into a nature journal
- pencil
- colored pencils

Optional

- camera and printer
- tape

Objectives

In this experiment, students will observe plants growing in their environment.

The objectives of this lesson are for students to:

- Explore the basic needs of plants—air, water, nutrients, temperature, and light.
- Examine different organisms in their environment.

Experiment

I. Think About It

Read this section of the *Laboratory Notebook* with your students.

Have the students think about plants they have already observed.

Explore open inquiry with questions such as the following:

- *What kinds of plants have you seen growing? Where were they growing?*

- *What kinds of plants have you eaten? Where do you think they came from?*

- *What features of plants have you observed?*
  (flowers, leaves, thorns etc.)

- *How do you think you can tell one kind of plant from another?*

- *How are plants different from rocks?*

- *Do you think you could put a plant in the soil anywhere and it would grow and be healthy? Why or why not?*

Have the students answer the questions in this section. There are no right answers.

II. Observe It

Read this section of the *Laboratory Notebook* with your students.

❶ For this experiment provide the students with a notebook or drawing pad that has blank pages. They will also use this notebook for the *Just For Fun* section, making it into a nature journal.

Take your students on a nature walk and help them observe the plants around them. They will need their notebook, a pencil, colored pencils, and a camera if one is available.

❷   Have the students choose two or more plants to study closely.

❸   Have them notice the size, shape, color, texture, and other features of the plants. Encourage them to draw the plants they've chosen. They don't need to be able to draw the plant accurately; attempting to draw it will cause them to look more closely at the plant and observe more details than if they just photograph it.

❹   Have the students observe the environment in which the plants are living, and in their notebook write details such as temperature, available water, soil conditions, and amount of sunlight or shade.

❺   If a camera is available, have them photograph the plants and tape the photos in the notebook next to their drawings.

III. What Did You Discover?

Have the students refer to their notes as they answer these questions. There are no right answers and their answers will depend on what they actually observed.

IV. Why?

Read this section of the *Laboratory Notebook* with your students.

Discuss any questions that might come up.

V. Just For Fun

Students are to make their notebook into a nature journal by observing the same plants over a period of several months and recording their observations. It would be helpful to set up a schedule for students to make observations on a regular basis.

The students may continue to observe the plants they chose for the *Observe It* section, or they may select different ones. However, once they have chosen the plants, they should observe the same plants for several months, looking for any changes they undergo.

Students may enjoy using their nature journal to make notes and drawings about other things they observe in the plants' environment, such as animals, birds, insects, rocks, etc. Encourage them to draw and write freely in their notebook.

BIOLOGY

Who Needs Light?

Materials Needed

- 2 small houseplants of the same kind and size
- 2 more small houseplants of the same kind and size
- water
- measuring cup
- closet or cardboard box
- colored pencils

Objectives

In this unit students will observe what happens to a plant if it does not get sunlight.

The objectives of this lesson are:

- For students to make careful observations and to compare a plant grown with sunlight to one grown without sunlight.
- To introduce the concept of using a *control*.

A *control* is a tool scientists use to compare the specific effect that making a change has on an experiment. By comparing the plant that stays in the sunlight (the control) to a plant that does not get sunlight (the unknown), students can better observe the effect that the absence of sunlight will have on the plant. Without a control, it can be hard to know for certain what caused the observed changes.

Experiment

I. Think About It

Read this section of the *Laboratory Notebook* with your students.

❶ Have the students think about what things plants need to have in order to live. Some of the basic things plants need are sunlight, air, water, minerals, etc.

❷ Have the students answer the question that asks what they think will happen if a plant does not get any sunlight. This may seem obvious to the students, but help them think about the details. Use questions such as:

- *What do you think will happen to the leaves if there is no sunlight?*

- *What color do you think the leaves will turn?*

- *What do you think the leaves will feel like after a few days without sunlight? Firm or soft?*

- *How many days do you think it will take for the plant that is without sunlight to show some problems?*

- *What do you think will happen first?*

- *What do you think will happen last?*

II. Observe It

Read this section of the *Laboratory Notebook* with your students.

❶-❷ Have the students look carefully at the two plants.

❸ Help them find words to describe their plants in detail.

Have them notice anything different between the plants.

❹ Have them label one plant "**A**" and the other plant "**B**."

Have the students draw their plants. Drawing helps students make more detailed observations.

This step sets up the first part of the experiment. It is important for students to record, in as much detail as possible, the substances and conditions present when an experiment begins. This way, the changes that occur during the experiment can be more easily tracked.

❺-❻ Have the students place the plant labeled **A** in a sunny place and the plant labeled **B** in a dark place. A dark closet would work well, but a cardboard box could also be used as long as it does not let in any light.

❼ Have the students think about what they might observe and then record their ideas.

❽ Guide the students in coming up with a schedule for watering the plants on a regular basis and help them decide how much water to use each time. Have them measure the water each time they water the plants.

Have the students draw what has happened to the plants after one week. Help them observe any differences.

Depending on the type of plant you have selected, it may be several weeks before a significant difference is observed. Have the students observe the plants weekly and record any changes they observe.

III. What Did You Discover?

Read this section of the *Laboratory Notebook* with your students.

Based on their actual observations, have the students answer the questions about what happened to the two plants. Have them write about any significant differences they observed.

BIOLOGY

IV. Why?

Read this section of the *Laboratory Notebook* with your students.

Discuss what happens when a plant does not get enough sunlight to be able to make its own food. Also discuss how using a control helped in comparing normal plant growth in sunlight to abnormal plant growth with no sunlight. Help the students understand that by using a control, they can make direct comparisons between plants that are subject to two different conditions—sunlight or no sunlight. Explain to them that a control helped them to determine specifically what effect sunlight, or the lack of it, had on the plants, since the amount of exposure to sunlight was the only factor that was different between the two plants—everything else should have stayed the same.

V. Just For Fun

In this experiment students will take two houseplants of the same kind and size and water one but not the other. Both plants should be placed near each other so all the parameters of the experiment are the same except for how much water the plants get.

Have the students review the experiment they performed in the *Observe It* section. Help them think about what modifications they need to make to come up with the steps for this experiment. Ask questions such as the following:

- *What do you think will happen to the plants?*

- *Which steps of the experiment will stay the same? Which steps will you change?*

- *After what length of time do you think you will notice a change in the plants?*

- *How frequently will you make observations?*

- *Do you think both plants should be kept near each other? Why or why not?*

Space is provided for beginning and ending observations. Students can choose to use additional paper to record more observations.

BIOLOGY

Experiment 8

Thirsty Flowers

Materials Needed

- 2-4 white carnations
- 1 or more other white flowers (rose, lily, etc.)
- 2-3 small jars
- food coloring
- water
- tape
- knife

- colored pencils

Optional

- magnifying glass

Objectives

In this unit students will observe how water travels through a plant stem and a flower.

The objectives of this lesson are to have students:

- Make careful observations about how plants use their stems for "drinking" water.
- Compare what they think will happen to the flower to what they actually observe.

Experiment

I. Think About It

Read this section of the *Laboratory Notebook* with your students.

❶ Have the students think about what will happen to the flower of a carnation if they put the stem in colored water. Help them be as specific as possible. Use questions such as:

- *What do you think will happen to the flower if you put the stem in blue water? Why?*

- *Do you think all of the petals will change color? Why or why not?*

- *Do you think only some of the petals will change color? Why or why not?*

- *Do you think none of the petals will change color? Why or why not?*

- *How do you think the petals may change color? From the end to the center? Or from the center to the end? Why?*

- *Do you think you will be able to see the colored water in the stem? Why or why not?*

- *Do you think the green color of the stem will color the flower? Why or why not?*

- *If you add yellow coloring to the blue water, will the flower turn blue and yellow? Or some other color? Why?*

❷ Have the students draw a picture of what they think will happen, showing details.

II. Observe It

Read this section of the *Laboratory Notebook* with your students.

❶ Have the students carefully observe and draw a white carnation. Help them examine any fine details they find interesting.

❷ Take the carnation and split it in two (or have the students cut it) lengthwise.

Have the students draw the inside of the stem and flower.

❸ Put some water in one of the jars. Add several drops of food coloring, using enough to deeply color the water. You may need to adjust the amounts of water and food coloring. Too much water and too few drops of food coloring will make the dye too dilute, and the coloring of the petals won't be very dark.

Have the students place a carnation in the jar. Make sure the end of the stem is fully submerged in the colored water. You may need to have the students tape the side of the stem to the jar or prop the flower so it does not come out of the water.

Have the students draw the "start" of the experiment. Have them note details, such as how the carnation is fixed to the jar or if it is tilted or how well the stem is submerged.

Have them observe the carnation for the next several minutes. As they note changes, have them draw the flower and record the number of minutes that have passed. Four boxes have been provided for recording these observations. It may take many minutes before the petals of the carnation are fully colored. Have the students make as many observations as possible. Help them pay attention to how the petals are being colored—from the top, side, bottom and so on.

❹ Cut the stem open, or have the students do this, and have them observe the inside. Guide their observations with questions such as:

- *Can you see the colored water traveling through the stem?*

- *Can you tell which part of the stem had the water traveling upward?*

- *Do you notice anything interesting about the stem? If so, describe it.*

III. What Did You Discover?

Read this section of the *Laboratory Notebook* with your students.

Have the students answer the questions about what they observed during this experiment. Help them think about the comparison between the way the carnation looked before and after it was put into the colored water.

IV. Why?

Read this section of the *Laboratory Notebook* with your students.

Discuss how the plant "drank" water from the jar. Tell the students that it is similar to how they drink liquid from a straw. When they put their mouth on the straw and suck in the air that is in the straw, liquid moves up from the bottom of their drink. A plant does essentially the same thing, except the "suction" comes from water evaporating from the leaves and petals.

If the students were able to observe differences inside the stem (such as parts of the stem that were light green, dark green, or white), explain to them that a stem has several different types of tissues. One of those tissues (called the xylem) draws water and nutrients up from the soil through the roots. Another type of tissue (called the phloem) pulls food back down through the stem to the roots and other parts of the plant. Tell them that only one type of tissue inside the stem draws the water up from the bottom of the jar, and the liquid will not drain back out again.

V. Just For Fun

Have students repeat the experiment using one or more different white flowers. Have them observe whether anything different happens.

BIOLOGY

Experiment 9

Growing Seeds

Materials Needed

- 1-2 small clear glass jars
- 2 or more dried beans (white, pinto, soldier, etc.)
- 2 or more dried beans of a different kind or 2 or more other seeds
- absorbent white paper
- scissors
- knife
- plastic wrap
- tape
- rubber band
- water

Optional

- magnifying glass

Objectives

In this unit students will observe how a seed grows into a plant.

The objectives of this lesson are for students to:

- Make careful observations about how a seed grows.
- Compare what they think will happen to what they actually observe.

Experiment

I. Think About It

Read this section of the *Laboratory Notebook* with your students.

❶ Have the students think about what will happen if they put a bean in a jar, add water, and let it sit for several days. Help them be as specific as possible. Direct their inquiry with questions such as:

- *What do you think will happen to a bean that gets water?*
- *Do you think the roots will come out first?*
- *Do you think the leaves will come out first?*
- *Do you think the bean will change color?*
- *What do you think might happen to the skin on the bean?*
- *How long do you think it will take for the bean to start to grow?*

Have them record their ideas.

❷ Have the students draw what they think will happen.

II. Observe It

Read this section of the *Laboratory Notebook* with your students.

❶ Have the students carefully observe and draw the outside of the bean. Help them examine any fine details they find interesting.

❷ Help the students split the bean lengthwise into two parts and then have them draw the inside of the bean, including any details they notice. A magnifying glass can be used while making the observations.

BIOLOGY

Guide the students in performing the following steps of the experiment.

❸   Take a clear glass jar and a piece of absorbent white paper. Cut a piece of the paper that is long enough to go all the way around the jar. Then wrap it around the inside of the jar.

❹   Place two dried beans between the paper and the jar. The paper should hold the beans against the side of the jar, but you may need to tape the beans to the jar if the paper doesn't hold them in place.

   Make sure the beans are not touching the bottom of the jar but are placed about 6-12 mm (1/4-1/2 inch) above the bottom.

❺   Pour some water in the bottom of the jar so that the water contacts the absorbent paper but not the beans. (**Note:** *The beans will rot if they are in the water.*)

❻   Place plastic wrap on top of the jar and fasten it with a rubber band to seal the jar and prevent evaporation of the water.

❼   Have the students draw the start of the experiment. Have them note details, such as how the beans are oriented in the jar—up, down, sideways, etc.

❽   Have them observe the beans as they grows into plants. Beans generally begin to germinate in 5-7 days. It may take several weeks for the bean plants to fully develop. Have the students record all of their observations as they watch the beans grow. Have them record observations for each bean, comparing any similarities and differences between the beans.

   Have the students check the water level frequently. It is important that the paper stay moist.

   Allow the beans to fully sprout. Both the roots and leaves should be clearly visible. Have the students note the direction in which the roots grow and the direction in which the leaves grow.

   If you wish to continue the experiment, you can have the students plant the seedlings and observe what happens. Do they become healthy plants? Why or why not?

III. What Did You Discover?

Read this section of the *Laboratory Notebook* with your students.

Have the students answer the questions about how the beans grew. Help them think about their observations and write summary statements about what they observed. Have them note whether or not the beans grew as they expected.

If the beans did not grow, guide the students in thinking about why this may have happened.

IV. Why?

Read this section of the *Laboratory Notebook* with your students.

Discuss the observations students made about how the beans grew. They should have noticed that the roots of the bean plant emerged first, followed by the leaves. They should also have observed that the roots grew down, toward the Earth, and the leaves grew up, toward the Sun.

Ask them how they think a plant "knows" in which direction to grow the roots and in which direction to grow the leaves. Explain that plant roots have molecules inside that tell roots to grow downward and that leaves have different molecules telling the leaves to grow upward toward the Sun.

V. Just For Fun

The students will repeat the experiment, this time using different seeds. They can use a different kind of dried bean seed, get seeds from the store, or save seeds from raw fruit or vegetables they eat. If they are saving their own seeds from food they eat, have them let the seeds dry before they begin the experiment to prevent the seeds from getting moldy. If they would like to experiment with several different seeds, they can put some different kinds of seeds in the same jar or use more than one jar.

Have the students record their observations. One box has been provided, and students can use additional pieces of paper if they choose.

BIOLOGY

Experiment 10

Lemon Energy

Materials Needed

- 3-5 large lemons
- knife
- 3-5 copper pennies older than 1982
- 3-5 galvanized (zinc coated) nails
- LED (Radio Shack #276-30700 [as of this writing])
- 4-6 pairs alligator clips
- plastic coated copper wire, .6-1.2 m (2-4 feet)
- wire clippers
- small Phillips screwdriver

[duct tape can be substituted for alligator clips]

Objectives

In this experiment students will explore the concept of stored chemical energy by making a battery from lemons.

The objectives of this lesson are to have students:

- Observe how the stored chemical energy in a lemon can power a small LED.
- Summarize their observations.

Experiment

I. Observe It

Read this section of the *Laboratory Notebook* with your students.

Assembling the Three-Lemon Battery Electric Circuit

❶ Help the students make two slits in each lemon to insert a penny and a galvanized (zinc coated) nail. Make sure the slits go past the lemon rind and into the fleshy part of the lemon. Have the students place a penny in one slit and a zinc nail in the other slit of each lemon. These are the battery "leads."

❷ Using the wire clippers, cut a piece of coated copper wire 15-20 cm (6-8 inches) in length, and strip off about 1.5 cm (one-half inch) of plastic coating from each end. Using the Phillips screwdriver, attach a red alligator clip to one end of the wire and a black alligator clip to the other end. Now connect the red alligator clip to the penny of one lemon (the positive lead) and the black alligator clip to the zinc nail of a second lemon (the negative lead). Repeat, connecting the third lemon. If you don't have alligator clips, you can use duct tape but it may not stick very well.

You should now have a series of lemons connected by wires with alternating penny and zinc leads. The ends of the two lemons on the outside of the series will have one end of a wire attached and one end free.

❸ The LED will be connected to the remaining unattached alligator clips or wire ends.

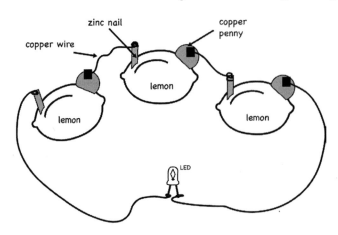

❹ When the students connect the LED to the free alligator clips or wires, the LED should light up. The light may be weak, so you may have to dim the room lights.

If the LED does not light up:

> Try switching the wire connections to the LED. It has positive and negative leads, and they may have been hooked up backwards. If the pennies are dull, clean them with a mild abrasive. Also, check the connections for all the pennies and zinc nails. Make sure all connections are secure, metal is touching metal, and the pennies and zinc nails are deep enough in the lemon to contact the fleshy part of the fruit. If the LED still won't light, try adding one or two more lemons to the circuit.

Use questions such as the following to help students make their observations.

- *Can you see any light coming from the LED?*

- *Are all the wires securely attached?*

- *Are the pennies and zinc nails in the soft part of the lemon?*

❺ Have the students record their observations in the space provided. They can write a description and/or draw a picture.

❻ Have the students remove one of the wires attached to one of the pennies. Guide their observations with questions such as the following:

- *Can you see any light coming from the LED?*

- *Reattach the wire. Does the light come back on?*

- *Remove the wire again. Does the light go off?*

❼ Have the students record their observations in the space provided. They can write a description and/or draw a picture.

❽-❾ Have the students reattach the wire to the penny and repeat steps ❻-❼, this time disconnecting a wire attached to a zinc nail. Use the same questions as in Step ❻.

In the chart provided, have the students summarize their observations. The expected results are as follows.

PHYSICS

Summarize Your Observations

Trial	What Happened?
All Wires Connected	The LED was illuminated
Penny Wire Disconnected	The LED was not illuminated
Zinc Wire Disconnected	The LED was not illuminated

II. Think About It

❶ Have the students think about their experiment and make any observations about the lemon battery circuit that seem important to them. Use questions such as the following to help them think about the experiment.

- *How easy or difficult was it to set up the lemon battery circuit?*

- *How bright was the LED when all the wires were connected?*

- *How did the LED respond when you removed one of the copper wires from either the penny or the zinc nail?*

❷ Have the students review Chapter 10 of the textbook which covers batteries and chemical energy. Discuss the concept that chemical reactions can produce electrical energy inside batteries. Help them relate this fact to the lemon battery.

❸ Have the students decide which statement is true. They should circle the statement "The LED will light up only when all the wires are connected."

❹ Have the students discuss any problems they may have encountered while doing the experiment. Some possible problems are:

- *Dull pennies (if the pennies are dull, there is an oxide layer on the outside of the metal which will prevent electrical contact).*

- *Lemon rinds that are too thick or too thin.*

- *Wires falling off.*

- *Tape not sticking.*

- *LED is defective.*

- *Too few lemons.*

Have a discussion with the students about how problems occur while doing experiments and that this is a normal part of doing science. Have the students discuss possible ways to fix the problems they encountered.

III. What Did You Discover?

Read this section of the *Laboratory Notebook* with your students.

With these questions, help the students think about their observations. There are no "right" answers to these questions, and it is important for students to write or discuss what they actually observed. Help them explore how the answers they got may be different from what they thought might happen.

Have the students compare what happened to the LED when all the wires were attached and what happened to the LED when they disconnected one or more of the wires.

IV. Why?

Read this section of the *Laboratory Notebook* with your students.

Have a discussion with the students about how they were able to use lemons as a battery. Also discuss with them how they created an electric circuit by connecting the lemon batteries together. In order to light an LED, at least three lemons are needed. Each lemon generates about 0.5 to 0.75 V of electric current, and an LED generally needs at least 2.0 V of electricity to illuminate. By combining the lemons together in a series, the voltage of each lemon is added to the others, and the total amount of electricity is enough to light the LED. Also discuss how when they disconnected one of the wires, the circuit was "broken," and the flow of electrons could no longer reach the LED.

V. Just For Fun

First, have the students make sure that all connections are secure, and then have them disconnect one of the wires from the LED. Have them hold the disconnected end of the LED wire in one hand and the end of the wire connected to a lemon in the other hand. Their body is now part of the electric circuit, conducting electricity and causing the LED to illuminate.

PHYSICS

Experiment 11

Sticky Balloons

Materials Needed

- 2-3 rubber balloons
- string or thread, at least 2 meters (6 feet) cut in half
- scissors
- different materials to rub the balloon on, such as:
  - cotton clothing
  - silk clothing
  - wool clothing
  - wooden surface
  - plaster wall
  - metal surface
  - leather surface

Objectives

In this experiment students will explore static electricity and how charges can transfer from one object to another.

The objectives of this lesson are for students to:

- Observe how an object can become charged.
- Observe how a charged object can generate an attractive force.

Experiment

I. Observe It

Read this section of the *Laboratory Notebook* with your students.

In this section students will perform a simple experiment to explore the transfer of static electric charges. They will explore the concept that different kinds of materials and surfaces will donate electrons to a rubber balloon. Rubber has a greater attraction for electrons than some other materials, such as wool, hair, silk, or fur and so will become charged when rubbed against these materials.

❶ Have the students blow up a rubber balloon. They will need to securely tie the end. Have them place the balloon on a wall. The balloon should not be charged and, unless it has picked up charges from being handled, it will simply fall off the wall. This step is the "control."

❷ Have the students rub the balloon in their hair.

❸ Have them pull the balloon away from their hair and observe whether the balloon pulls their hair. If the balloon pulls their hair, it has attracted electrons from the hair and can be tested for an electric charge. If the balloon does not pull the hair, have the students rub the balloon in their hair again. In humid climates this experiment may not work very well since the moisture in the air prevents the charges from sticking to the balloon.

❹ Have the students "test" for the electric charge on the balloon by placing it on a wall.

❺ Help the students record their observations in the space provided. Use the following questions (also listed in the *Laboratory Notebook*) to guide their inquiry.

PHYSICS

- Does the balloon stick?

- How long does the balloon stick? 1 second? 2 seconds? 10 seconds? Longer than a minute?

- Does the balloon move around or stay still?

- What happens if you blow gently on the balloon? Does it stay stuck or does it fall off?
  (The balloon should stick to the wall if there have been enough charges transferred from the hair to the balloon.)

Hair

(Example. Answers may vary.)

The balloon got really charged with my hair. It pulled my hair

to the sides. The balloon really stuck to the wall, and even

when I blew on the balloon it didn't move. It stayed on the wall

longer than a minute.

Have the students draw what they observed.

❻ Have the students repeat Steps ❷-❺, rubbing the balloon on other materials. To discharge the balloon, wipe the outside of the balloon with a moist paper towel. Once the balloon is discharged, have the students rub the balloon against a different material or surface to recharge it. For each material have the students test the balloon to see how well it sticks to the wall. Have them record their observations, including how long the balloon stays on the wall each time.

II. Think About It

Read this section of the *Laboratory Notebook* with your students.

❶ Help the students think about their experiment and make any observations about the different materials they used to create charge on the balloon. You can use questions such as the following to guide the discussion:

- *Was it easy to get the balloon to carry a charge?*

- *What problems did you have in getting the balloon charged?*

- *Did some materials work better than other materials?*

- *How well did the balloon stick to the wall?*

- *Do you think using a wall to stick the balloon to is a good way to test for charge? Why or why not?*

❷ Chapter 11 of the textbook covers electrons, charges, and force. Review this chapter with the students and help them understand that an electron carries a charge. The experimental results they observe are due to charges jumping from one object (hair, silk cloth, fur) to another object (the balloon).

❸ The wall test is used to determine how much charge the balloon collected from the various materials or surfaces. This is a qualitative estimation, and there may be some surfaces or materials that do not differ much from each other. You can help the students guess which materials may have had more charge by having them observe the length of time the balloon stayed on the wall. The longer the balloon stuck to the wall, the more charge it picked up from the material.

Have the students review their observations and create a chart listing the materials or surfaces used and how they affected the balloon from **Most Charge** to **Least Charge**.

III. What Did You Discover?

Read this section of the *Laboratory Notebook* with your students.

With these questions, help the students think about their observations. There are no "right" answers to these questions, and it is important for the students to write or discuss what they actually observed. Help them explore how the answer they got may be different from what they thought might happen.

Help the students explore how easy or difficult this experiment was to perform. Also, help them think about why they may get different results from the different materials used. Some materials may donate more charge to the balloon than other materials. The wall test is a way to determine how "much" charge the balloon collected. Help the students explore whether this is a good way to determine charge on the balloon. Ask the students if they can think of any other way to test the charges.

IV. Why?

Read this section of the *Laboratory Notebook* with your students.

Have a discussion about rubber being a material that easily picks up electrons from other surfaces. Discuss the "control" that was performed at the beginning of the experiment. By blowing up the balloon and applying it to the wall before they rubbed it in their hair, they got a feel for how the balloon behaves when it is uncharged.

Depending on the weather conditions in your hometown, this experiment was either easy to perform or did not work well. If the air is very dry, like in arid parts of the country, then static electric charges are easily created. If the air is more humid, then static charges are more difficult to create. Therefore, the results may vary depending on where you live.

V. Just For Fun

Have the students take two balloons and tie a piece of string or thread on one end of each balloon and then tie the other ends of the two pieces of string together.

Have the students drape the tied balloons over a shower rod, or help them fix the balloons to a doorway. Since the balloons are not charged, they should float free, touching each other but not touching any other surface.

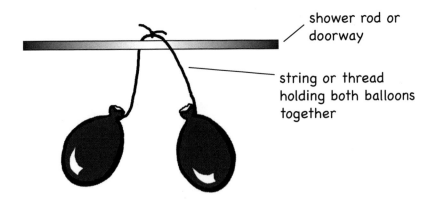

shower rod or doorway

string or thread holding both balloons together

Help the students take the balloons and rub them in their hair. Have them gently let go of the balloons and watch what happens. If both balloons are charged, they will repel each other. Explain to the students that when things have the same kind of charge, they repel each other. When they have opposite charges, they attract each other.

Have the student record their results.

PHYSICS

Experiment 12

Moving Electrons

Materials Needed

- lemon battery supplies (see Experiment 10)
- suggested test materials:
  Styrofoam
  plastic block
  cotton ball
  nickel coin
  metal paper clip
  plastic paper clip
- glass of water
- table salt, 15 ml (1 Tbsp.)

Objectives

In this experiment students will explore moving electric current and observe how insulating materials resist the flow of electrons through them.

The objectives of this lesson are to have students:

- Observe the effect of insulating materials on electric flow.
- Organize results in two different ways.

Experiment

I. Observe It

Read this section of the *Laboratory Notebook* with your students.

Students will perform a simple experiment to explore how different materials conduct electricity. They will test both metals and nonmetals and compare their results.

❶ Help the students assemble the lemon battery from Experiment 10. Complete the setup exactly as in Experiment 10, and confirm that the LED is illuminated.

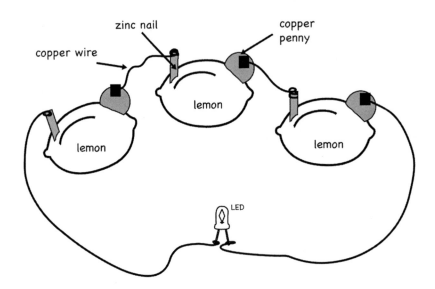

❷ Have the students cut one of the wires that connects two of the lemons. Tell them to observe the LED as they cut the wire.

❸ To reconnect the wire, help the students strip off about 1 cm (1/2 inch) of plastic coating from the ends of the cut wire. Have them reconnect the metal ends by briefly touching them together to make the LED illuminate, or they may want to twist the ends around each other. Make sure that metal is connected to metal and that a metal end has not been wrapped around the plastic coating. Have the students observe the LED and write and/or draw their observations in the space provided.

❹ Have the students disconnect the cut ends of the wire once again. Have them stick each metal end in a piece of Styrofoam. The ends should not touch each other, and the piece of Styrofoam will complete the connection. Have the students observe the LED and write and/or draw their observations.

❺ Have the students remove the Styrofoam and reconnect the wires as they did in Step ❸. Have them observe what happens to the LED and then write and/or draw their observations in the space provided.

❻ Have the students repeat Steps ❹-❺ using the other items on the materials list. If some of the materials are not available, other materials can be substituted. Have the students test a few insulators (plastics, cotton, wool, etc.) and a few conductors (metals).

❼ Have the students summarize their results in the table provided. If an item has been substituted, have them mark out the listed item and write in the new item. There should be a mixture of LED "ON" and LED "OFF" results as shown in the chart below.

Item	LED
Start—wires connected	ON
Wires apart	OFF
Wires connected	ON
Styrofoam	OFF
Wires connected	ON
Plastic block	OFF
Cotton Ball	OFF
Nickel coin	ON
Metal paper clip	ON
Plastic paper clip	OFF

PHYSICS

II. Think About It

Read this section of the *Laboratory Notebook* with your students.

❶ Help the students think about their experiment and make any additional observations about the different materials they used to test the flow of electrons. Use questions such as the following to guide their inquiry.

> • *Did you notice a difference in how the LED responded when different materials were used as part of the connection?*
>
> • *What problems did you have connecting the wires to the different materials?*
>
> • *Do you think some of the problems you had affected your results?*
>
> • *Do you think the LED would light up if you used bigger (or smaller) pieces of plastic, Styrofoam, or cotton?*
>
> • *Do you think the LED would be brighter if you used larger pieces of metal?*

❷ Review with the students Chapter 12 of the *Student Textbook* which covers moving charges. Help the students understand that in order for electrons to flow through a material, the electrons have to be free to jump from one atom to the next. Have them think about the fact that insulators are insulators all the way down to their atoms and that using a larger piece of plastic or Styrofoam will not make an insulator conduct electricity.

❸ Have the students use the chart provided to organize the materials they tested. This exercise allows the students to explore alternative ways of examining scientific data. Explain to the students that scientists will often reorganize their data in ways that help them understand different aspects of the experiment. Here the students will see that those items that illuminated the LED are grouped together as conductors and those items that did not illuminate the LED are grouped together as insulators.

III. What Did You Discover?

Read this section of the *Laboratory Notebook* with your students.

Have the students think about their observations and then answer the questions. There are no "right" answers to these questions, and it is important for the students to write or discuss what they actually observed. Help them explore how the answer they got may be different from what they thought might happen.

PHYSICS

Help the students examine the types of materials that illuminated or did not illuminate the LED. It will be clear from the data that plastics, Styrofoam, and cotton are like each other in that they are all materials that do not allow the flow of electrons (i.e., are insulators). It will also be clear from the data that coins, metal paper clips, and other metals do allow the flow of electrons (i.e., are conductors). Help the students use the language of "conductors" and "insulators" to formulate their answers.

IV. Why?

Read this section of the *Laboratory Notebook* with your students.

In this experiment students explored how conductors and insulators behave differently. Conductors, such as metals, allow electrons to flow and permit the illumination of the LED. Insulators do not allow electrons to flow and do not permit the illumination of the LED. Explain that the experiment they set up is a good way to determine which materials conduct electricity and which materials do not. Also explain that there are some materials that allow a restricted electron flow. These materials are called resistors.

V. Just For Fun

Have the students take the ends of the cut wire and place them in a glass of water. Depending on the type of tap water you are using, the LED may or may not illuminate.

Have the students add 15 ml (1 Tbsp.) of table salt to the water and stir until it is completely dissolved. The LED should now illuminate when the ends of the wires are placed in the saltwater.

Discuss with the students how this experiment showed that water and saltwater are conductors because they allow electrons to flow through them.

PHYSICS

Experiment 13

Magnet Poles

Materials Needed

- two bar magnets with the poles labeled "N" and "S"

Objectives

In this experiment students will explore the nature of magnetic poles.

The objectives of this lesson are to have students:

- Observe how magnetic poles behave when placed close together.
- Make a scientific conclusion based on their observations.

Experiment

I. Observe It

Read this section of the *Laboratory Notebook* with your students.

In this section the students will perform a simple experiment to explore how magnetic poles attract and repel each other.

❶ Have the students place the bar magnets several inches apart on a table with the "N" on one magnet facing the "N" on the other. The magnets need to be far enough apart that there are no repulsive forces interacting.

❷ Have the students gently push the end of one magnet closer to the end of the other magnet. Have them go slowly so they can observe what happens with the stationary magnet. At some point the stationary magnet will change. It will move away from the magnet that is being pushed (the stationary magnet will be repelled). At this point don't discuss the "north" and "south" poles of the magnets, just focus the students' attention on observing what is happening to the magnets. Help them write and/or draw their observations in the space provided. This is *Trial 1*.

❸ Again have the students place the two magnets on the table several inches apart. Have them reverse the magnetic pole of one magnet by flipping the magnet over so the "S" pole of this magnet is facing the "N" pole of the other magnet.

❹ Have the students gently push the magnets together as they did in Step ❷. This time the magnets should be attracted to each other and may even snap together. This is *Trial 2*.

❺ Have the students repeat Steps ❶-❹ several times.

Help the students make observations for each trial. Guide inquiry with questions such as the following:

- How close can you bring the magnets together before something changes?

- What happens when you bring the magnets together slowly?

- What happens when you bring the magnets together quickly?

- Do the magnets behave the same way each time you bring two "N"s together?

- Do the magnets behave the same way each time you bring an "N" and an "S" together?

❻ Help the students summarize their results in the chart provided.

Trial	N–N or N–S	Together or Apart
Trial 1	N–N	Apart
Trial 2	N–S	Together
Trial 3	N–N	Apart
Trial 4	N–S	Together
Trial 5	N–N	Apart
Trial 6	N–S	Together
Trial 7	N–N	Apart
Trial 8	N–S	Together
Trial 9	N–N	Apart
Trial 10	N–S	Together

II. Think About It

Read this section of the *Laboratory Notebook* with your students.

❶ Help the students think about the magnets and the two different ways the magnets were positioned as they were pushed together.

❷ Review with the students Chapter 13 of the *Student Textbook* which covers magnetic poles. Have a discussion about magnets having two opposite poles that are called "north" and "south." Have the students look at the magnets and observe the "N" and "S" letters.

PHYSICS

❸ Have the students look at the results they gathered in the previous table. In this table they recorded N-N or N-S notations along with the observations they made of whether the magnets moved together or apart. Have the students cover up the column that has the N-N and N-S notations, and have them look only at the column that says whether or not the magnets moved together or apart. Using this information, have the students guess which trials had the "same" poles and which trials had the "opposite" poles, and have them record their answers in the table provided.

Trial	Same or Opposite
Trial 1	Same
Trial 2	Opposite
Trial 3	Same
Trial 4	Opposite
Trial 5	Same
Trial 6	Opposite
Trial 7	Same
Trial 8	Opposite
Trial 9	Same
Trial 10	Opposite

❹ Have the students uncover the hidden column and compare the two tables. Have them note whether or not their answers in the second table match the second column of the previous table.

In this exercise they are using their scientific results to make a guess, predicting which poles were the same and which were opposite based only on their observations.

III. What Did You Discover?

Read this section of the *Laboratory Notebook* with your students.

With these questions, help the students think about their observations. There are no "right" answers to these questions, and it is important for the students to write or discuss what they actually observed. Help them explore how the answer they got may be different from what they thought might happen.

PHYSICS

Help the students summarize their results. Have them discuss what happened when they pushed the two "N" poles together and when they pushed the "N" and "S" poles together. Ask them if, based on this experiment, they could take two magnets that were not labeled "N" and "S" and predict which poles were the same and which were opposite. Help them understand that they do not need to know "N" and "S" but only "same" and "opposite." Remind them that opposite poles attract and similar poles repel and that they can predict which poles are the same or opposite by whether or not the magnets are attracting each other or repelling each other.

IV. Why?

Read this section of the *Laboratory Notebook* with your students.

In this experiment students explored the magnetic poles of magnets. These poles interact and produce either an attractive or repulsive force. The students took two magnets, and by bringing the ends of the magnets together they explored how the poles interact with each other.

The students also did several "trials" to explore the magnetic poles. By repeating the steps of the experiment several times, the students could "play" with how the magnets interact with each other, making observations for each trial. Scientists generally do more than one trial when they are conducting experiments. By playing with the experiments, scientists discover things that they might have missed if they only did the experiment once.

V. Just For Fun

Have the students test the magnets with different household items. Have them touch a magnet to a number of different objects, both metal and nonmetal, to find out which materials will interact with the magnet.

Have them record their results in the table provided.

PHYSICS

Experiment 14

How Fast Is Water?

Materials Needed

- 3 Styrofoam cups: 355 ml (12 oz.) size
- about 240 ml (1 cup) each:*
  sand
  pebbles
  small rocks
- 3 containers for collecting sand, pebbles, and small rocks
- garden trowel or small shovel
- pencil
- 1-2 measuring cups
- water

  * student-collected or purchased from a place that sells aquarium supplies

Optional

- stopwatch or clock with second hand

Just For Fun

- enough dirt, pebbles, rocks, water, etc. to make a mud city

Objectives

This experiment introduces students to the concepts of porosity and groundwater. Students will explore how the porosity of a material determines how quickly water can flow through it and how porosity affects the absorption of rainwater into the soil.

The objectives of this lesson are for students to:

- Observe how different materials have different porosities.
- Observe how the porosity of a material affects how water flows through it.

Experiment

I. Think About It

Read this section of the *Laboratory Notebook* with your students.

Have the students think about what happens to rain after it falls to the ground. Guide open inquiry with questions such as:

- *Where do you think the water in lakes and rivers comes from? Why?*

- *What do you think happens to rain after it falls on the land? Where does it go? Why?*

- *Do you think there is a difference between rain falling on the ground and snow falling? Why or why not? What is different?*

- *Do you think you would see a difference between rain that falls on rocks and rain that falls in a garden? If so, what differences would you see?*

- *Why do you think rain sometimes makes mud puddles and sometimes does not?*

- *What do you think keeps the water in a lake?*

- *Do you think plants and animals could live on land where there is no rain or snow? Why or why not?*

GEOLOGY

II. Observe It

Read this section of the *Laboratory Notebook* with your students.

If possible, have the students collect the sand, pebbles, and small rocks to be used in this experiment.

❶ Have the students use a pencil to poke a hole in the bottom of each Styrofoam cup. The holes should be about the same size.

❷-❹ Have the students measure about 240 milliliters (1 cup) each of the sand, pebbles, and small rocks and pour each into its own cup.

❺-❼ The students are to pour 120 milliliters (4 ounces) of water into each of the three cups and observe how fast the water flows through each material.

The objective of this part of the experiment is for students to compare the relative speed at which water travels through the three different materials. Students can measure the length of time by using a stopwatch or a clock with a second hand, or simply by observing how quickly or slowly water runs through each material compared to the others.

Guide the students to observe how much of the water comes out the bottom of each cup and whether the amount varies by material. They can check it visually or catch the water in a measuring cup.

Have them record their observations.

III. What Did You Discover?

Read this section of the *Laboratory Notebook* with your students.

Have the students review their notes from the *Observe It* section and answer the questions based on their observations. There are no right or wrong answers.

IV. Why?

Read this section of the *Laboratory Notebook* with your students.

Discuss any questions that might come up.

GEOLOGY

V. Just For Fun

By building a mud city with a landscape that has rivers and a lake, students can begin to explore how surface water and groundwater operate. Students can use various natural materials to build their city — dirt, sand, pebbles, rocks, water, etc. They can try different things like slowing the speed of the river water by using pebbles or rocks or speeding it up by making a smooth riverbed and a higher elevation for the source of the river. A dam can be built to contain the lake water. Allow the students to play with the materials and see what they can discover on their own.

Help the students think about how they want to build their city by asking questions such as:

- *Where do you think the city part should go? Do you think the city would have a wall around it?*

- *Where will the rivers be?*

- *How will you make the water flow in the rivers?*

- *Where will the river water end up?*

- *Can you make the river water move faster or slower?*

- *Where will you put the lake?*

- *How will you keep the water in the lake?*

- *If the water doesn't stay in the lake, where do you think it will go?*

Students can make paper boats to float in their rivers and lake. If they'd like, they can draw their mud city.

GEOLOGY

Experiment 15

What Do You See?

Materials Needed

- pencil
- colored pencils

Objectives

In this experiment students will explore their environment by walking around their yard and neighborhood, observing plants, insects, animals, and people and any activities they are involved in.

The objectives of this lesson are for students to:

- Observe living things and the environment they live in.
- Observe some of the factors that create that environment.

Experiment

I. Think About It

Read this section of the *Laboratory Notebook* with your students.

By having students think about the living things they will observe when they go out into their environment, they will begin to understand that the living things and the resources within an environment are interconnected and that environments differ from region to region.

Guide open inquiry with questions such as the following.

- *If you walk around outside, what animals, birds, and bugs do you think you will see?*

- *What will the animals, birds, and bugs be eating? Why? What will they be doing? Why?*

- *Where do you think different animals, birds, and bugs sleep? Why?*

- *Do you think all living things sleep? Why or why not?*

- *Do you think you will see the same plants everywhere you look? Why or why not?*

- *Can you think of any animals that need to live near or in water?*

- *Do you think you will see any people? What do you think they'll be doing? Why?*

- *Do you think you'll see places where people can live and places where they can't live? Why or why not?*

II. Observe It

Read this section of the *Laboratory Notebook* with your students.

Have the students walk around their yard and neighborhood, observing what's around them. Encourage them to take their time and to look at things close up as well as from a distance. In addition to observing living things, they can also be guided to notice the weather, the presence or absence of water, the soil, and other factors that go into creating a particular environment.

Guide open inquiry with questions such as:

- *What are the animals, birds, and bugs doing?*

- *What are they eating?*

- *Are they moving around? How?*

- *Are any of them sleeping? Where?*

- *Is the weather making any difference in what the animals and birds are doing? Why or why not?*

- *Are the plants different in different areas? Why or why not?*

- *Are there different animals in different areas? Why or why not?*

- If there is a watery place that can be observed—*Can you tell if there are plants or animals living in the water? What are they doing? What do they eat? Could they live without the water? Why?*

- *Do you see any people? What are they doing? Why?*

Have the students bring their *Laboratory Notebook*, a pencil, and colored pencils with them on their walk so they can make drawings and notes as they make their observations.

The blank line in each heading on the observation pages can be filled in with the location of the observations or some other identification.

III. What Did You Discover?

Read this section of the *Laboratory Notebook* with your students.

Have the students refer to the drawings and notes they made while on their walk. Answers are based on their observations, and there are no right or wrong answers to these questions.

IV. Why?

Read this section of the *Laboratory Notebook* with your students.

Discuss any questions that might come up.

V. Just For Fun

Since the Kepler spacecraft was launched in 2009, scientists have discovered that many stars have planets orbiting them. Those planets that exist outside our solar system are called exoplanets.

The Circumstellar Habitable Zone is the area of a solar system where an Earth-like planet would be at the right temperature to have liquid water and therefore might have the right conditions for life as we know it to exist. Kepler-62e is a recently discovered Earth-like exoplanet in a Circumstellar Habitable Zone and may have conditions suitable for life, although that is not yet known as of this writing.

In this experiment students are to use their imagination to think about what life might be like on another planet. There are no right or wrong answers. Encourage students to go where their imagination takes them, even if their ideas are improbable, wild, and fanciful.

GEOLOGY

Moving Iron

Materials Needed

- 2 bar magnets (narrow magnets work best)
- small, flat-bottomed, clear plastic box (big enough for 2 magnets to fit underneath with some space around them)
- corn syrup
- iron filings, about 5 ml (1 teaspoon) (see Experiment section for how students can collect iron filings— or iron filings may be purchased: www.hometrainingtools.com)

Optional

- tape
- 2 plastic bags for collecting iron filings

Objectives

In this experiment students will "see" the magnetic field around a magnet.

The objectives of this lesson are for students to:

- Visualize the magnetic field surrounding a magnet.
- Observe how a magnetic field can influence objects that are attracted to magnetic force.

Experiment

Students can gather iron filings themselves by putting a magnet in a plastic bag and dragging the bag through some dirt. Iron that is in the dirt will collect on the outside of the bag. Place the bag containing the magnet inside another plastic bag and then remove the magnet from the inner bag. The iron filings will fall into the outer bag. Repeat several times until about 5 ml (1 teaspoon) of iron filings has been collected.

Alternatively, iron filings can be purchased.

I. Think About It

Read this section of the *Laboratory Notebook* with your students.

Have the students think about magnets and what they have learned about them. Guide open inquiry with questions such as:

- *Do you think everything made of metal can be a magnet? Why or why not?*

- *If you look at a magnet, do you think you will see the magnetic field that surrounds it? Why or why not?*

- *If you were an astronaut in space, do you think you could see Earth's magnetic field? Why or why not?*

- *Do you think magnets can be useful? Why or why not? What could they be used for?*

- *What do you think would happen if you pushed two magnets together matching their north poles? Matching the north pole of one with the south pole of the other? Why?*

GEOLOGY

II. Observe It

Read this section of the *Laboratory Notebook* with your students.

❶ Have the students pour corn syrup into the box. There needs to be enough syrup that the iron filings can move around freely. A layer of about 6 millimeters (1/4") works well.

❷ Have the students place the box on top of the magnet so that the magnet is centered. The box needs to be flat (not tipped).

   Optional: The magnet can be taped to the bottom of the box.

❸ Have the students pour the iron filings on the syrup, being careful not to breathe them in.

❹ After about 30 minutes the students will be able to observe how the magnetic field of the magnet has affected the iron filings. The corn syrup is not affected by the magnetic field, but the iron filings will have aligned to the magnetic forces.

❺ Have students record what they observe.

III. What Did You Discover?

Read this section of the *Laboratory Notebook* with your students.

Have the students answer the questions based on their observations. Guide open inquiry with questions such as:

- *Do you think the magnetic field of a magnet will always stay the same? Why or why not?*

- *Why do think you had to wait 30 minutes before recording the pattern made by the iron filings?*

- *Do you think a different magnet would make the iron filings move into a similar pattern? A different pattern? Why or why not?*

- *Do you think Earth's magnetic field might have a pattern that is similar to that of the bar magnet? Why or why not?*

GEOLOGY

IV. Why?

Read this section of the *Laboratory Notebook* with your students.

Discuss any questions that might come up.

V. Just For Fun

❶ Have the students move the box so the magnet is repositioned. Depending on the interest of the student, the magnet can be repositioned more than once. Students will need to wait about 30 minutes each time before making their observations.

Have them record the resulting pattern and observe whether repositioning the magnet changed the pattern of the magnetic field.

❷-❸ In this part of the experiment students will place two magnets under the box to see if adding a magnet changes the magnetic field. Depending on their interest, they can reposition the magnets more than once.

Have them record the resulting pattern(s) and observe whether using two magnets changed the pattern of the magnetic field and also if the pattern was different when the two magnets were moved to new positions.

What Do You Need?

Materials Needed

- seeds (student selected)
- a garden bed or containers and potting soil
- tools for tending plants
- herb seeds or small herb plants (student selected)

This experiment is done over the course of several weeks.

Objectives

The experiment will take place over the course of several weeks. Students will grow a plant and observe how all the parts of the Earth are interdependent and necessary for plant life.

The objectives of this lesson are for students to:

- Observe factors that are necessary for a plant to grow and be healthy.
- Observe how all the parts of the Earth work together to provide everything needed for a plant to grow.

Experiment

I. Think About It

Read this section of the *Laboratory Notebook* with your students.

Have the students think about the role each part of the Earth plays in the growth of a plant. Guide open inquiry with questions such as:

> - *Do you think the atmosphere contains anything that a plant needs in order to grow? Why or why not?*
>
> - *Do you think plants can live without water? Why or why not?*
>
> - *How do you think plants get water from the hydrosphere?*
>
> - *Do you think plants need anything from the biosphere? Why or why not?*
>
> - *Where do you think soil comes from? Why?*
>
> - *Do you think plants could live if there were no magnetosphere? Why or why not?*

II. Observe It

Read this section of the *Laboratory Notebook* with your students.

❶ Help the students choose a plant to grow from seed. For the purposes of this experiment, a plant likely to be eaten by bugs would be a good choice, and one that produces vegetables that can be eaten could be fun to grow. If there's enough space, students might like to grow more than one type of plant.

❷ Have the students plant several seeds in case some of them don't germinate. They may need to thin out some of the plants if too many seeds sprout. Seeds may be planted in a garden bed or in a container in a sunny spot. It's best to have the container outside where the plant can interact with the weather, bugs, etc. If it isn't possible to have the plant outdoors, have the students think about what factors might be different for a plant that is indoors and what advantages and disadvantages an indoor plant might have.

❸ Help the students remember to water the seeds after they're planted and check the soil moisture daily until the seeds sprout. Then have them check the soil on a regular basis. Have them notice in what ways the weather affects the soil.

❹ Students are to write and/or draw any observations they make after they plant the seeds and as they water them. Boxes are provided in the workbook.

❺-❼ Have the students make frequent observations of their plant. Use questions such as the following to guide them in making observations about how different parts, or spheres, of the Earth are affecting their plant.

- **Biosphere:**
  Do you think any other living things are affecting your plant? What are they and what effect are they having?
  (For example, are bugs or rabbits eating the plant, worms enriching the soil, a cat rolling against the plant?)
  What things are you doing that affect the plant?

- **Atmosphere:**
  Do you think wind is affecting your plant? How?
  Is it too sunny, too cloudy, too rainy—or not enough of any of these? How can you tell?

- **Hydrosphere:**
  Is the plant getting enough water from rain (also involves the atmosphere)? Where do you think the water from the hose or spigot comes from? How and why?
  Can you tell when the plant needs more water? Why or why not?

- **Geosphere:**
  Do you think the soil affects how often the plant needs water (also involves the hydrosphere)?
  What does it look like the soil is made of?

- **Magnetosphere:**
  Do you think the magnetosphere is letting the right amount of the Sun's energy reach the plant? How can you tell?

Have the students check the growth and health of their plant frequently and write and/or draw what they observe.

III. What Did You Discover?

Read this section of the *Laboratory Notebook* with your students.

Have the students review their observations and use them to answer the questions. Discuss any questions that may arise. There are no right or wrong answers.

IV. Why?

Read this section of the *Laboratory Notebook* with your students.

Discuss any questions that might come up.

V. Just For Fun

Students are to grow an herb garden of several plants. It can be in a garden bed or in containers outdoors or indoors.

Help the students select herbs to grow. They can research herbs in the library or online or go to a nursery to look at plants. Starting with either seeds or small plants is suitable for this experiment. Students might choose herbs they know they like or ones that look pretty or have intriguing names. Or they may come up with their own selection criteria.

Discuss possible locations for the herb garden and have the students plant the seeds or repot the plants, if needed,

As the herbs are growing, have the students make observations about how the plants grow and what is needed to keep them healthy. The students can decide when to harvest leaves and observe how this affects the plants. If herbs suitable for making tea are grown, students can try cutting off some branches of the herb and hanging them upside down from a string tied around the stems. Once the leaves are dry, they can be put in hot water to make tea. Students can also use this method to dry herbs for later use in foods, storing the dried leaves in airtight containers.

Students who like doing research might enjoy discovering which plants have flowers that are edible and then grow and eat some of these.

GEOLOGY

Experiment 18

Modeling a Galaxy

Materials Needed

- student-selected materials to make a model of a galaxy, such as colored modeling clay, Styrofoam balls, tennis balls, marbles, sand, candies, etc.
- cardboard or poster board, .3-1 meter (1-3 feet) on each side

Optional

- colored pencils or markers
- camera and printer

Objectives

In this experiment, students will explore model building and its limitations.

The objectives of this lesson are to have students explore how:

- Models help scientists ask questions.
- Scientists use different kinds of investigation, such as model building.

Experiment

I. Think About It

Read this section of the *Laboratory Notebook* with your students.

Explore open inquiry with questions such as the following:

- *How many houses do you think are in our neighborhood?*

- *How many neighborhoods do you think are within walking distance?*

- *How many neighborhoods have you visited?*

- *Do you think some neighborhoods look different from others? Why or why not?*

- *What else is in our neighborhood?*

- *Do you think a galaxy has neighborhoods? Why or why not?*

- *Do you think if you went from one part of a galaxy to another both parts would look the same? Why or why not?*

- *How many solar systems do you think are in a galaxy? Do all galaxies have the same number of solar systems? Why or why not?*

II. Observe It

Read this section of the *Laboratory Notebook* with your students.

❶ Have the students think about the different objects present in a galaxy and which ones they would like to represent in their model. Mentioned in the *Student Textbook* are solar systems (stars and planets), black holes, dust, and gases. Help the students select materials to make a model of a galaxy. Their galaxy can be as simple or as elaborate as they choose. The size of the model will be limited by the size of the cardboard or poster board provided as well as by the size of the various materials selected.

Use questions such as the following to help students select the materials to use for representing the various parts of the galaxy and to think about the model design.

- *Which parts of a galaxy do you want to have in your model? Why?*

- *How might you show different neighborhoods in your galaxy model?*

- *What material(s) would you use to represent the stars? Why?*

- *Do you want all the stars in your model to be the same size? Why or why or not?*

- *What material would you use to represent the planets? Why?*

- *Do you want all the planets in your galaxy model to be the same size and color? Why or why or not?*

- *What material might you use to represent dust in the galaxy?*

- *What do you think a black hole would look like in your model?*

Have the students list the objects they want to represent in their model and the materials they will use.

❷ Provide the cardboard or poster board for the students and have them spread out their model making materials next to it or along an edge.

❸ Guide the students in designing and drawing their galaxy model by having them think about where the solar systems and any other objects might go, how many of them they will create, and how far apart and how big they will be. Have them think about the placement of different objects in terms of creating neighborhoods. There are no right answers and they can add as many solar systems and other objects as they like. Help them notice the limitations of their model.

❹ Have the students build the model galaxy, referring to the design they created. Have them construct at least one solar system with at least one Sun and 3 or more planets. Several smaller solar systems can be constructed or drawn on the cardboard. Solar systems may be any size. Have them add any other objects they want to represent.

❺ Have the students record any features of their model that they think are unique or unexpected. Or they might record the features they like best.

❻ Help the students photograph their model and print a copy to tape in the *Laboratory Notebook*, or they can draw it or describe it in writing.

ASTRONOMY

III. What Did You Discover?

Read this section of the *Laboratory Notebook* with your students.

Have the students answer the questions. There are no right answers and their answers will depend on what they actually observed..

IV. Why?

Read this section of the *Laboratory Notebook* with your students.

Discuss any questions that might come up.

V. Just For Fun

Encourage students to use their imagination freely to think about what it might be like to go to the end of the universe. Do they think there's an end? If so, what might it look like? What could be seen along the way there? Have them record their ideas by drawing and/or writing in the space provided.

See the Milky Way

Materials Needed

- colored pencils
- a dark, moonless night sky far away from city lights

Optional

- computer with internet access
- pictures of cities

Objectives

In this unit, students will use the model of a city to help them think about and observe the Milky Way Galaxy.

The objectives of this lesson are for students to:

- Observe a city to see that it typically has more buildings at its center than it does at the edges.
- Compare how observations made about a city's structure and organization can be used to better understand that of a galaxy.

Experiment

I. Think About It

Read this section of the *Laboratory Notebook* with your students.

Have students think about the density of buildings in the center of a city compared to their density on the edges of a city. Most cities have a greater number of close together buildings in the center and fewer, more widely spaced buildings on the outskirts. Have students think about how their observations of a city's structure can be used to model the structure of a galaxy.

Encourage open inquiry with questions such as the following.

- *How many buildings are in the downtown section of a city?*
- *How many buildings are at the edges of a city?*
- *Are the buildings in the downtown area typically taller/bigger or shorter/smaller than the buildings at the edges?*
- *Are the buildings in the downtown area typically closer together or farther apart than the buildings at the edges?*
- *How might the center of a galaxy like ours be compared to the center of a city?*
- *How might the edges of a city or the suburbs be compared to the edges of a galaxy?*
- *Do all cities have the same shape? Do all galaxies have the same shape? Why or why not?*
- *Do you think cities stay the same size? Do you think galaxies stay the same size? Why or why not?*

ASTRONOMY

II. Observe It

Read this section of the *Laboratory Notebook* with your students.

To observe the Milky Way, choose a clear, moonless night far away from city lights. Allow some time for students' eyes to adjust to the dark and have them look with their eyes only.

For students in the Northern Hemisphere, the best time to view the Milky Way is in the late summer and early winter when the Milky Way will look brighter. In the Southern Hemisphere winter is the best time for viewing.

**Best times and dates
for viewing the Milky Way in the Northern Hemisphere**

Summer Milky Way Times & Dates		*Winter Milky Way Times & Dates*	
8 PM	October 14	8 PM	February 10
9 PM	September 29	9 PM	January 25
10 PM	September 14	10 PM	January 10
11 PM	August 30	11 PM	December 26
12 AM	August 15	12 AM	December 12
1 AM	July 31	1 AM	November 26
2 AM	July 16	2 AM	November 11

**Best times and dates
for viewing the Milky Way in the Southern Hemisphere**

Summer Milky Way Times & Dates		*Winter Milky Way Dates & Times*	
8 PM	April 8	8 PM	August 28
9 PM	March 24	9 PM	August 12
10 PM	March 9	10 PM	July 28
11 PM	February 21	11 PM	July 12
12 AM	February 5	12 AM	June 27
1 AM	January 22	1 AM	June 13
2 AM	January 7	2 AM	May 29

III. What Did You Discover?

Read this section of the *Laboratory Notebook* with your students.

Help students think about their observations while answering these questions. There are no "right" answers to these questions, and it is important for the students to write or discuss what they actually observed.

ASTRONOMY

IV. Why?

Read this section of the *Laboratory Notebook* with your students.

The Milky Way Galaxy contains a vast number of stars. The majority of stars in the galaxy reside in a flat disk, and there is a concentration of stars near the center. Our solar system is located within the flat disk of stars. Because Earth is located toward the outer edge of the galaxy, we can see a multitude of stars as we look through the galaxy toward the center. The band of light and stars we call the Milky Way appears as a band because we are looking edge-on through the disk of stars toward the galaxy center.

Artist's Rendition of Milky Way Galaxy as seen from outside
and looking toward the center   Credit: NASA/UMass/Caltech

V. Just For Fun

Help your students set up Google Earth:

❶ Go to http://earth.google.com and click "Download Google Earth."

❷ Click "Agree and Download."

❸ Once the file has been downloaded, follow the directions to install the program.

❹ Open the Google Earth program on your computer.

❺ Set up Google Earth in Sky mode by clicking on the planet symbol in the top menu bar and selecting "Sky" from the drop down menu. Or, at the top of the page, click "View," then "Explore," and then select "Sky."

❻ Type "Milky Way" into the search box. Once the Milky Way appears, you can zoom in and out using the +/- bar.

Students can also go to Google Images (www.google.com/imghp) and search on Milky Way Galaxy to find some beautiful photos and illustrations of the galaxy.

If your students are interested, you can have them use Google Earth to take a look at some of the planets.

ASTRONOMY

How Do Galaxies Get Their Shape?

Materials Needed

- 2 bar magnets
- iron filings, purchased* or student collected (see Chapter 16)
- shallow, flat-bottomed plastic container (or a plastic box top or large plastic jar lid)
- corn syrup
- plastic wrap
- Jell-O or other gelatin and items needed to make it
- assorted fruit cut in pieces and/or berries

Optional

- cardboard box

* As of this writing, available from Home Science Tools: http://www.hometrainingtools.com Item #CH-IRON

Objectives

In this unit students will model the forces that help shape galaxies.

The objectives of this lesson are for students to:

- Use model building as a way to explore elements of science that can't be directly tested.
- Observe how forces move objects.

Experiment

I. Think About It

Read this section of the *Laboratory Notebook* with your students.

Have the students think about how galaxies might form. Encourage open inquiry with questions such as the following:

- *How do you think a galaxy might form?*
- *What do you think holds the stars and planets together?*
- *Do you think galaxies can have any kind of shape? Why or why not?*
- *Do you think galaxies can change their shape? Why or why not?*
- *What do you think might happen if two galaxies collided?*

II. Observe It

Read this section of the *Laboratory Notebook* with your students.

❶ Provide the students with a shallow, flat-bottomed plastic container, and help them pour corn syrup into it until the syrup is just below the rim of the container.

❷ Help the students pour iron filings into the corn syrup. Iron filings can irritate eyes and lungs, so they should be handled carefully.

❸ The students should now cover the container with plastic wrap to prevent spilling.

❹ The container can be placed on a table that has a top thin enough for the magnets to work when they are being moved under it; or you can place a cardboard box on its side with the open end toward the student and the container on top; or the container can be held for the student. There should be enough space under the container for both of the student's hands to fit and be able to move.

ASTRONOMY

Have students place the magnets underneath the plastic container and observe the movement of the iron filings. The corn syrup will create drag on the iron filings, slowing the movement. Encourage students to be patient while observing the filings.

❺ Have the students take one of the magnets and create a swirling pattern with it. The swirling motion will simulate how a spiral galaxy might form. Have the students notice that the source of the force (the magnet) moves and that this creates movement in the iron filings.

❻ Now have them take a magnet in each hand and create opposite swirling patterns.

❼ Have the students bring the magnets together to see what happens.

❽ Allow the students some time to "play" with the magnets and iron filings. Slowing down to "play" with science is an essential part of scientific inquiry. Encourage the students to bring the magnets together, pull them apart, swirl them together, swirl them separately, swirl them in opposite directions, and so on. How many different shapes can they create?

III. What Did You Discover?

Read this section of the *Laboratory Notebook* with your students.

In answering these questions, help the students think about their observations. There are no "right" answers to these questions, and it is important for students to write or discuss what they actually observed.

IV. Why?

Read this section of the *Laboratory Notebook* with your students.

Gravity is a force that acts between any two objects that have mass, pulling them toward each other. Magnetism depends on the arrangement of electrons in objects and can either pull objects toward each other or repel them, moving them away from each other. All objects are affected by gravity, but not all objects are affected by magnetism.

Even though magnetic forces are different from gravitational forces, in this experiment the magnetic action is pulling on the iron filings so the force appears to be like that of gravity. Therefore, magnets can be used to simulate how gravity works to pull and shape galaxies. The magnetic forces act on the iron filings, and when the magnets are moved, the iron filings are pulled along. In a similar way stars pull on smaller objects such as planets, comets, and asteroids. Stars also pull on neighboring stars and together they shape a galaxy.

V. Just For Fun

Help your students set up a Jell-O galaxy. Any type of fruit can be used to represent planets and stars. Help your students cut the fruit into small pieces and arrange the fruit or swirl it in the Jell-O mixture to create an elliptical galaxy, a spiral galaxy, or an irregular galaxy. Encourage them to think about whether or not other shapes of galaxies might be possible.

Making a Comet

Materials Needed

- small plastic pail that will fit in freezer
- water
- dirt
- small stones
- dry ice (available at most grocery stores)
- heavy gloves or oven mitts
- freezer

If dry ice comes in a block:

- safety goggles
- mallet or hammer
- cloth or paper grocery bag

Objectives

In this unit, students will model a comet.

The objectives of this lesson are for students to:

- Observe how a mixture of water, dirt, and rocks can be frozen to form a model comet.
- Explore how changing the ratio of water:dirt:rocks changes a comet.

Experiment

I. Think About It

Read this section of the *Laboratory Notebook* with your students.

Encourage open inquiry with the following questions.

- *How big do you think a real comet is? Why?*
- *Do you think a real comet is mostly ice or mostly dirt or mostly rock?*
- *How much ice do you think is needed to hold rocks and dirt together?*
- *Do you think a real comet breaks apart, or does it vaporize as it nears the Sun? Could it do both? Why or why not?*

II. Observe It

Read this section of the *Laboratory Notebook* with your students.

In this experiment students will make a model comet from water, dirt and rocks. Help the students make careful observations about the comet they are creating.

❶ Have students collect some dirt and small stones.

❷ Have students pour dirt and stones into a small pail and add enough water to cover, leaving several inches between the top of the pail and the water. As the water freezes it will expand.

❸ Place the pail in the freezer and allow enough time for the water to freeze completely.

❹ Have the students tap the frozen comet model out of the pail. They may need to run a little warm water over the outside of the pail to get the frozen mixture to release.

❺ Encourage the students to observe the frozen comet model carefully and draw or write about any details they notice.

❻ Students will now observe their comet model as it melts. Guide them in making their observations. Do the rocks fall away in large chunks? What happens to the dirt? What else do they notice? Have them record their observations.

❼-❽ Have the students repeat the experiment first with more water and then more dirt. By carefully recording their observations, they will be able to compare how a comet containing lots of ice acts compared to a comet made mostly of dirt.

III. What Did You Discover?

Read this section of the *Laboratory Notebook* with your students.

In answering these questions, help the students think about their observations. There are no right answers to these questions, and it is important for the students to write or discuss what they actually observed.

IV. Why?

Read this section of the *Laboratory Notebook* with your students.

A real comet is a large chunk of ice and rock that might look similar to the model comet the students have created. However, the ice in real comets vaporizes rather than melts. This experiment demonstrates what happens to a comet as it loses its ice, although it does not replicate the exact method by which the ice is lost. Explain to the students that astronomers performing experiments are not always able to duplicate the exact conditions that exist in space.

Comets also contain frozen gases such as carbon dioxide and carbon monoxide. The tails of comets will be different colors depending on the kinds of gases that are vaporizing from the comet.

V. Just For Fun

Using dry ice will create a more realistic comet. Dry ice is made of frozen carbon dioxide, and it vaporizes rather than melts. This is fun for the students to watch.

Dry ice is very cold and can burn the skin with minimal contact. Please **use extreme caution** when building a comet with dry ice, and don't let bare skin touch the dry ice. Wear heavy gloves or oven mitts.

If the dry ice is in a solid block, it will need to be broken into small pieces. You can place it in a cloth or paper grocery bag and hit it with a mallet or hammer until it is broken up. (You may want to wear goggles during this process.) Then the pieces can be added to the water, dirt, rock mixture.

ASTRONOMY

Experiment 22

All Science

Materials Needed

- library or internet access

Optional

- 1 or more old toys to take apart to look for computer chips

Objectives

In this experiment, students will explore how scientists share information and how this shared information has driven the creation of the computer. Students will also explore how the computer is being used in a multitude of modern devices.

The objectives of this lesson are for students to:

- Explore how science and technology make contributions to people and their lives.
- Learn about how people use scientific inquiry to make discoveries and invent new technologies.

Experiment

I. Think About It

Read this section of the *Laboratory Notebook* with your students.

In this experiment students will look at one technological invention—the computer. They will explore how science contributed to the invention of the computer and how the computer contributes to our every day lives.

Help students think about what a computer is made of, including plastic (from chemistry), metals (from chemistry), electric currents (from physics), software (from computer science and mathematics), etc.

Ask questions to guide students' exploration of one or more components of the computer in great detail. For example:

- *Think of the keyboard. What is a keyboard made of?*

- *What type of plastic is used? Is it hard plastic or soft plastic? Why?*

- *How is the keyboard designed? What does it do?*

- *What if you could only type with your toes. How might a keyboard for toes be designed?*

- *How is a keyboard connected to the computer?*

- *What are the wires made of?*

- *And so on.*

Have students answer the questions in this section.

II. Observe It

Read this section of the *Laboratory Notebook* with your students.

Help students explore how computers are used in a variety of everyday activities. Toys, games, cell phones, TVs, DVD players, bicycles, and cars are all products that may utilize a computer or computer chip.

Over the course of a week, have students observe and record all the different items they discover that contain a computer or computer chip.

III. What Did You Discover?

Read this section of the *Laboratory Notebook* with your students.

Have students answer the questions. Again, there are no right answers.

IV. Why?

Read this section of the *Laboratory Notebook* with your students.

Discuss any questions that might come up.

V. Just For Fun

Have students use the library or internet to research what a computer chip is and what functions it can perform. If the student has any old toys they can disassemble, allow them to take the toy apart to see if it has a computer chip. Many toys with moving parts or lights contain a computer chip or integrated circuit.

More REAL SCIENCE-4-KIDS Books
by Rebecca W. Keller, PhD

Building Blocks Series yearlong study program — each Student Textbook has accompanying Laboratory Notebook, Teacher's Manual, Lesson Plan, Study Notebook, Quizzes, and Graphics Package

Exploring Science Book K (Activity Book)
Exploring Science Book 1
Exploring Science Book 2
Exploring Science Book 3
Exploring Science Book 4
Exploring Science Book 5
Exploring Science Book 6
Exploring Science Book 7
Exploring Science Book 8

Focus On Series unit study program — each title has a Student Textbook with accompanying Laboratory Notebook, Teacher's Manual, Lesson Plan, Study Notebook, Quizzes, and Graphics Package

Focus On Elementary Chemistry
Focus On Elementary Biology
Focus On Elementary Physics
Focus On Elementary Geology
Focus On Elementary Astronomy

Focus On Middle School Chemistry
Focus On Middle School Biology
Focus On Middle School Physics
Focus On Middle School Geology
Focus On Middle School Astronomy

Focus On High School Chemistry

Super Simple Science Experiments

21 Super Simple Chemistry Experiments
21 Super Simple Biology Experiments
21 Super Simple Physics Experiments
21 Super Simple Geology Experiments
21 Super Simple Astronomy Experiments
101 Super Simple Science Experiments

Note: A few titles may still be in production.

Gravitas Publications Inc.
www.gravitaspublications.com
www.realscience4kids.com